T0313208

MicroRNA

Perspectives in Health and Diseases

MicroRNA

Perspectives in Health and Diseases

Edited by
Jaishree Paul
Rohini Muthuswami

CRC Press
Taylor & Francis Group
Boca Raton London New York

CRC Press is an imprint of the
Taylor & Francis Group, an **informa** business

CRC Press
Taylor & Francis Group
6000 Broken Sound Parkway NW, Suite 300
Boca Raton, FL 33487-2742

Printed on acid-free paper

International Standard Book Number-13: 978-1-1380-5483-7 (Hardback)

Library of Congress Cataloging-in-Publication Data

Names: Paul, Jaishree, editor.
Title: MicroRNA : perspectives in health and diseases / [edited by] Jaishree Paul, Rohini Muthuswami.
Description: Boca Raton : Taylor & Francis, 2018. | "A CRC title, part of the Taylor & Francis imprint, a member of the Taylor & Francis Group, the academic division of T&F Informa plc." | Includes bibliographical references and index.
Identifiers: LCCN 2017047023| ISBN 9781138054837 (hardback : alk. paper) | ISBN 9781315166391 (ebook)
Subjects: LCSH: MicroRNA--Health aspects. | Small interfering RNA.
Classification: LCC QP623.5.S63 M5256 2018 | DDC 572.8/8--dc23
LC record available at https://lccn.loc.gov/2017047023

Contents

Foreword

Ever since the classic work of Jacob and Monod demonstrated that the lac repressor is a protein, gene regulation has been understood almost entirely on the basis of protein regulators. Antisense RNA-mediated regulation was reported and well-documented in bacteria from the 1970s, but RNA was never considered a serious contender for gene regulation. Even the discovery in 1993 of the lin-4 microRNA in nematodes was not considered to be of general significance. From these beginnings, the eventual realization that small RNAs are all-pervasive modulators of gene regulatory networks has been nothing short of a revolution, especially since proteins seemed to be doing this role to perfection. The major class of regulatory noncoding RNAs is the microRNAs (miRNAs), which post-transcriptionally regulate their specific target genes, and are now known to regulate a variety of developmental and physiological processes. Not surprisingly, miRNA levels have been linked to many human disorders, including pathogenic infections, and undoubtedly more such linkages will continue to emerge.

With the surge of research in miRNA regulatory networks operating in various human diseases, this has been the subject of a number of authoritative volumes in the past. The present volume is unique in that it documents some of the latest research in this area in Indian labs. The first chapter describes the basic biology of miRNA biogenesis, to bring into perspective the known mechanisms for miRNA generation. The second chapter deals with computational methods to predict miRNA genes and their target genes, and to understand the mechanism of miRNA function by theoretical modeling of gene regulatory networks. The subsequent chapters discuss the involvement of miRNAs in a variety of human diseases and disease processes, including glioblastoma, neurogenesis and neurodegeneration, multiple sclerosis, inflammatory bowel disease, parasitic infections, hepatocellular carcinoma, Alzheimer's disease, and metabolic disorders. The authors highlight the miRNA-based therapeutic strategies, and the use of miRNAs as biomarkers for monitoring disease progression. Pathogens can exploit host miRNAs, as well as express their own miRNAs to promote their survival and evade host immune response. In other diseases, the expression of different miRNA classes is associated with disease progression and can be used for early detection and prognosis. The replacement or inhibition of miRNAs could also be used for potential therapy. Obviously, the pattern of expression of disease-related miRNAs could be specific to different ethnic populations, which makes the studies with Indian patients essential and important. This volume will be invaluable to students, researchers, and clinicians and would hopefully boost further research with miRNAs to harvest the potential of these molecules for the benefit of patients.

Sudha Bhattacharya
Professor
School of Environmental Sciences
Jawaharlal Nehru University, New Delhi

Preface

The discovery of microRNA (miRNA) has revolutionized our understanding of how the expression of a gene is regulated. The traditional view of gene expression, either by transcription factors or by translational regulators, has been upended by these small RNA molecules that bind to mRNA, resulting either in its degradation or sequestration such that protein expression is altered. Today, the miRNAs are acknowledged as major contributors to cellular homeostasis. The role of miRNA in disease manifestation is also becoming increasingly evident, and these molecules are being considered as biomarkers for various diseases. They are also being studied as a potential drug target for therapeutic purposes.

The present book, divided into nine chapters, delves into the world of miRNA and provides a glimpse of the research work being done by researchers in India. We begin with Chapter 1, an introduction to miRNA and how the biogenesis of these molecules is regulated both at transcriptional and post-transcriptional levels. The regulation of miRNA is important for cellular function, and dysregulation of miRNA has been linked to diseases such as cancer.

We then introduce in Chapter 2 the *in silico* tools available to identify miRNA as well as to predict the mRNA target for a given miRNA. The chapter also introduces theoretical models available to understand the miRNA regulatory network.

The next seven chapters are focused on understanding the role of miRNA in human diseases. The miRNA profile is known to be altered when protozoans infect us. Chapter 3 delineates these changes and how they impact the establishment of infection in humans by manipulating the host defense mechanisms. Chapter 4 focuses on multiple sclerosis (MS), a disease whose etiology is not well-understood. As the authors show, the role of miRNA in MS has been established without doubt and using these molecules as drug targets is one method being considered for the treatment of this disease. Chapter 5 focuses on the role of miRNA in inflammatory bowel disease (IBD). The IBD is a perfect example of the role of both environment and genetic makeup in disease manifestation. There are two forms of IBD: ulcerative colitis and Crohn's disease. The authors explore how miRNA influences the development of IBD in individuals. They also explain the differential expression of miRNA in the two forms of IBD and finally, explore the miRNA that causes IBD in Indian population.

Chapters 6 and 7 focus on neurogenesis and neurodegenerative disorders. Alzheimer's disease results in memory loss in elderly patients, and the authors report the role of miRNA in Alzheimer's disease. They report the role of miRNA in six broad areas of Alzheimer's disease. Chapter 7 explains how miRNA helps in neurogenesis process and how dysregulation results in neurodegeneration.

Finally, we turn our attention to cancer. Chapters 8 and 9 focus on the role of miRNA in tumorigenesis and cancer progression. The miRNAs are divided into two classes: oncogenic miRNAs and tumor suppressor miRNAs. The authors discuss the changes that occur in miRNA expression in hepatocellular carcinoma and the potential use of these miRNAs, both as biomarkers and drug targets. Glioblastoma multiforme is the most aggressive type of cancer of the central nervous system, and the authors of the final chapter provide a comprehensive overview of the interplay of the miRNA and the regulatory pathways in the regulation of this invasive cancer. They also provide a detailed account on the possible use of miRNA-based combination therapy in combating glioblastoma multiforme.

We hope that this book will provide readers a comprehensive view on the current developments in the area of miRNA research involving its potential role in the pathogenesis of a disease.

We express our sincere appreciation to all the authors who have enriched the book with their valuable contributions. We express our sincere thanks to all our students and colleagues for their cooperation and timely help. Last but not the least, we are grateful to our publisher CRC, Taylor & Francis for providing us the opportunity to carry out this project and their editorial assistance.

Editors

Prof. Jaishree Paul is currently working as a scientist at Jawaharlal Nehru University, New Delhi. She is currently involved in exploring and understanding the basic mechanisms of inflammatory bowel disease, particularly ulcerative colitis. She has more than 25 years of teaching experience and has published more than 50 research papers. She has solely authored a book entitled *Handbook of Disease Causing Microbes* published by Ane books and CRC press and has contributed chapters in six books with Elsevier and with local publishers of repute. She has been invited as a speaker at many international conferences. She has been awarded Visitor's Research award by the President of India on 14th March, 2016.

Dr. Rohini Muthuswami is currently working as a scientist at Jawaharlal Nehru University, New Delhi. She is currently involved in knowing the epigenetic mechanisms that regulate the cellular processes. She has published 15 research papers and has written a book on *Exploring the Biological World* with National Book Trust, India and has also contributed a chapter on Chromatin Remodeling in Gene and Engineering. She has four patents in her name as well.

Contributors

Sneha Anand
Molecular Parasitology Laboratory
School of Life Sciences
Jawaharlal Nehru University
New Delhi, India

Ganapathy Ashok
Department of Molecular Microbiology
School of Biotechnology
Madurai Kamaraj University
Madurai, Tamil Nadu, India

Sanghamitra Bandyopadhyay
Machine Intelligence Unit
Indian Statistical Institute
Kolkata, India

Alok Bhattacharya
School of Life Sciences
Jawaharlal Nehru University
New Delhi, India

Malay Bhattacharyya
Department of Information Technology
Indian Institute of Engineering Science
    and Technology
Shibpur, Botanic Garden
West Bengal, India

Muthiah Chellappandian
Division of Biopesticides and
    Environmental Toxicology
Sri Paramakalyani Centre for
    Excellence in Environmental
    Sciences (SPKCES)
Manonmaniam Sundaranar University
Tamil Nadu, India

Villianur Ibrahim Hairul Islam
Biological Sciences Department
College of Science
King Faisal University
Hofuf Ahsaa, Saudi Arabia

Ritu Kulshreshtha
Department of Biochemical
    Engineering and Biotechnology
Indian Institute of Technology Delhi
New Delhi, India

Venugopal Senthil Kumar
Pondicherry Centre for Biological
    Sciences and Educational Trust
Pondicherry, India

Rentala Madhubala
Molecular Parasitology Laboratory
School of Life Sciences
Jawaharlal Nehru University
New Delhi, India

Amit Kumar Mishra
South Asian University
Akbar Bhawan
New Delhi, India

Pamchui Muiwo
School of Life Sciences
Jawaharlal Nehru University
New Delhi, India

Rohini Muthuswami
Chromatin Remodeling Laboratory
School of Life Sciences
Jawaharlal Nehru University
New Delhi, India

Priyatama Pandey
School of Computational and
    Integrative Sciences
Jawaharlal Nehru University
New Delhi, India

Kishor Pant
South Asian University
Akbar Bhawan
New Delhi, India

Ketki Patne
Chromatin Remodeling Laboratory
School of Life Sciences
Jawaharlal Nehru University
New Delhi, India

Jaishree Paul
School of Life Sciences
Jawaharlal Nehru University
New Delhi, India

Beena Pillai
CSIR—Institute of Genomics and
    Integrative Biology (IGIB)
Sukhdev Vihar
New Delhi, India

Mayuresh Anant Sarangdhar
CSIR—Institute of Genomics and
    Integrative Biology (IGIB)
Sukhdev Vihar
New Delhi, India

Subramanian Saravanan
Department of Clinical & Translational
    Science
Creighton University School of
    Medicine
Omaha, Nebraska

Krishnaraj Thirugnanasambantham
Pondicherry Centre for Biological
    Sciences and Educational Trust
Pondicherry, India

Swati Valmiki
School of Life Sciences
Jawaharlal Nehru University
New Delhi, India

Senthil Kumar Venugopal
South Asian University
Akbar Bhawan
New Delhi, India

Omkar Vinchure
Department of Biochemical
    Engineering and Biotechnology
Indian Institute of Technology Delhi
New Delhi, India

1 Controlling the Biogenesis of the Smallest Regulators

Ketki Patne and Rohini Muthuswami

CONTENTS

Introduction

miRNAs were first discovered in 1993 by Victor Ambros in *C. elegans* and now are known to be found in plants, animals and some viruses (Lee et al., 2007, 1993). These small (~22 base) single-stranded regulatory RNAs are involved in most cellular processes, including cell growth, development, differentiation, homeostasis and apoptosis (Bartel, 2004). miRNAs adhere with Argonaute proteins to form the RNA-induced silencing complex (RISC) that binds to the target mRNA and mediates post-transcriptional gene regulation by translational inhibition or mRNA degradation (Yang and Lai, 2011). The importance of regulation of miRNAs can be inferred from the observation that alteration in miRNAome (the entire complement of miRNAs expressed by a cell, tissue or organism) of a cell is linked to many diseases including cardiovascular diseases, neurodevelopmental diseases, autoimmune diseases and cancer (Ardekani and Naeini, 2010).

The biogenesis of most of the miRNAs involves two key processing steps, but some miRNAs, which bypass this canonical pathway, are also known (Daugaard and Hansen, 2017).

The first step of canonical miRNA biogenesis pathway (see Figure 1.1) involves transcription of the miRNA gene by RNA polymerase II/III to form a primary transcript (primary miRNA or pri-miRNA) that can be several kilobases long and

1

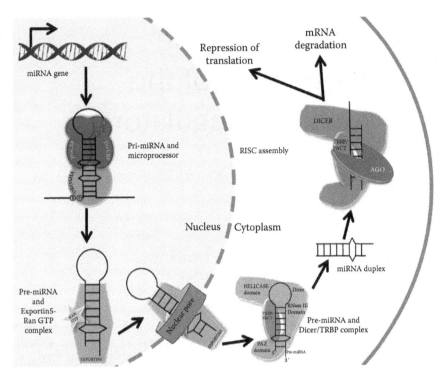

FIGURE 1.1 miRNA biogenesis pathway. Mature miRNAs are produced in a cell by the action of two sets of processing enzymes – Drosha in the nucleus and Dicer in the cytoplasm. Drosha forms the microprocessor complex with DGCR8 and cleaves the pri-miRNA into pre-miRNA. These pre-miRNAs are acted upon by Dicer in cytoplasm to form the miRNA duplex. The thermodynamically less stable strand is then loaded in the RISC complex where it guides Ago2 to its mRNA target causing translational repression or mRNA degradation.

contains many stem-loop structures, each of which resembles a different precursor miRNA (pre-miRNA) (Borchert et al., 2006; Lee et al., 2004; Winter et al., 2009). The pri-miRNA undergoes the first processing step in the nucleus where it is cleaved by the microprocessor complex to form the pre-miRNA (Denli et al., 2004; Gregory et al., 2004). The microprocessor complex is composed of two key proteins, Drosha and DGCR8, and several additional proteins (Denli et al., 2004; Gregory et al., 2004; Han et al., 2004; Landthaler et al., 2004; Lee et al., 2003). During the processing of pri-miRNA, DGCR8 recognizes and binds to the stem and loop region and helps Drosha to cleave at the base of this stem-loop structure to form a 70 base long pre-miRNA (Lee et al., 2003; Nguyen et al., 2015).

The pre-miRNA is then transported to the cytoplasm by Exportin5-RanGTP complex, where it goes through the final processing step (Bohnsack et al., 2004; Lund et al., 2004; Yi et al., 2003). The cytoplasmic processing step includes cleavage by Dicer-TRBP-PACT complex to form the miRNA duplex (Bernstein et al., 2001; Hutvágner et al., 2001; Ketting et al., 2001; Wilson et al., 2015). This complex then gets linked with the Argonaute protein where the thermodynamically more stable strand of the miRNA duplex is degraded, and the less stable strand forms the RNA induced silencing (RISC) complex (Chendrimada et al., 2005; Gregory et al., 2005; Khvorova et al., 2003). The miRNA then guides the RISC complex to the target mRNA where the Argonaute protein exerts its function, leading to translational inhibition or mRNA degradation (Gregory et al., 2005).

miRNAs are important regulators of most cellular processes. Therefore, the levels of miRNAs need to be tightly regulated, specialized to the cell type and depending on the cell state. The miRNAome of a cell is maintained by controlling the various steps of miRNA biogenesis pathway.

1.1 TRANSCRIPTION OF miRNA GENES

miRNA genes can be located within the intronic or exonic region of a protein-coding gene or a non-coding transcription unit where they are transcribed by the promoter of the host gene (Kim et al., 2009). Moreover, they can also be found in the intergenic regions where they are guided by their own promoter (Kim et al., 2009). A miRNA gene can be transcribed either independently or as a cluster along with other miRNA genes (Altuvia et al., 2005). Clustered miRNAs are those that are transcribed as a single primary transcript, which is then cleaved by the microprocessor complex, to give rise to individual pre-miRNA molecules.

The location of a miRNA gene decides its way of transcription and whether it will be transcribed by RNA polymerase II or III. Intragenic miRNAs are transcribed by RNA polymerase II under the influence of promoter of the host gene (Lee et al., 2004). miRNA genes located within or near the tRNA genes, mammalian-wide interspersed repeat (MWIR) sequences or Alu elements, are transcribed by RNA polymerase III, and can be controlled transcriptionally by similar mechanisms as these repeats (Borchert et al., 2006). miRNA genes that are clustered together are transcribed as a single primary transcript (Baskerville and Bartel, 2005) while non-clustered miRNA genes are transcribed individually by their own promoter (Lee et al., 2004).

1.2 FACTORS REGULATING THE TRANSCRIPTION OF miRNA GENES

Tissue-specific miRNAome expression plays fundamental roles in distinctive features of cell types and tissue-specific pathways. Tissue-specific gene expression patterns are modulated by various transcription factors that are also known to regulate protein-coding genes. Many factors bind to the promoter of miRNA genes and regulate their transcription. The proto-oncogene c-Myc is one such factor which binds to the promoters of miR-17-92 cluster and causes its transcriptional upregulation. The miRNAs miR-17-5p and miR-20a of this cluster then regulate cell cycle progression by targeting the cell cycle promoter E2F1 (O'Donnell et al., 2005). The binding of c-Myc to the promoter region may also cause inhibition of transcription, as happens with some tumor suppressor miRNA genes including miR-15a, miR-29, miR-34 and let-7 family (Chang et al., 2008).

miRNA biosynthesis is also regulated by some auto-regulatory feedback and feed forward loops (Tsang et al., 2007). For example, transcription factors E2F1, E2F2 and E2F3 can also activate the miR-17-92 cluster; and miR-20a of the cluster then targets the mRNA of E2F1, E2F2 and E2F3, creating an auto-regulatory feedback loop (Sylvestre et al., 2007). Similarly, the expression of miR-145 can be stimulated by tumor suppressor p53, and in turn, miR-145 activates p53, creating a feed forward loop in breast cancer cell line MCF-7 (Spizzo et al., 2010). Another p53-stimulated miRNA, miR-34a, targets SIRT-1 and leads to the activation of p53 (Yamakuchi et al., 2008).

miRNAs also take part in signaling pathways in the cell. A good example of a feed forward loop in a signaling cascade is the Ras/MAPK pathway, where, KRAS causes transcriptional inactivation of miR-143/145 cluster through transcription factor RREB1 (Ras-responsive element-binding protein1), and thus, promotes tumorigenesis. In turn, miR-143/145 target KRAS and RREB1, generating a feed forward regulatory loop (Kent et al., 2010). Besides a negative feed forward loop, Ras signaling also encompasses a positive feed forward loop, where it transcriptionally activates miR-21 via the transcription factor AP-1. miR-21 then upregulates AP-1 levels by targeting its inhibitor PDCD4 (Talotta et al., 2009).

The TGF-β pathway is known to cause transcriptional activation of miR-216a and miR-217 genes; these miRNAs then target PTEN (phosphatase and tensin homologue) and thus, activate the Akt kinase, boosting cell survival in glomerular mesenglialcells (Kato et al., 2009).

The p53 DNA damage pathway has also been shown to regulate the biogenesis of many miRNAs. For example, p53 is known to upregulate the cell cycle inhibitory miRNAs by increasing their transcription as well as by enhancing the processing steps (Liu et al., 2017). Other than miR-143 and miR-34a, it can also bind to the gene promoters of other miR-34 family members and cause their transcriptional upregulation (Chang et al., 2007; Corney et al., 2007). The miR-34 family members then target and downregulate the cell cycle protein CDK6 resulting in termination of the cell cycle (Suzuki et al., 2009).

Epigenetic modification of miRNA genes is also a good way to regulate transcription from these genes. DNA methylation and histone modification are two important epigenetic modifications found in miRNA genes. Epigenetic modifications play a vital role in many cancers where the tumor suppressor miRNA genes are found to be hypermethylated (Lujambio et al., 2008). Similarly, alteration in acetylation levels of histones has been linked to tumorigenesis and metastasis (Glozak and Seto, 2007; Gray and Teh, 2001).

1.3 PROCESSING OF THE PRIMARY miRNA TRANSCRIPT

The processing of pri-miRNA transcripts is done by the microprocessor complex which consists of two key proteins, Drosha and DGCR8 (DiGeorge syndrome critical region 8), and many accessory proteins (Gregory et al., 2004). Of all the microprocessor complex proteins, Drosha and DGCR8 form the core complex which is sufficient for the recognition and processing of pri-miRNAs. The accessory proteins help in conditioning and fine-tuning the processing step for specific miRNAs. The pri-miRNAs are around 70 nucleotides long and have stem-loop secondary structures (see Figure 1.2). The recognition of pri-miRNAs by Drosha-DGCR8 complex occurs primarily by its secondary structure. Earlier it was thought that DGCR8 recognizes the pri-miRNA with the proper secondary structure and aligns Drosha near

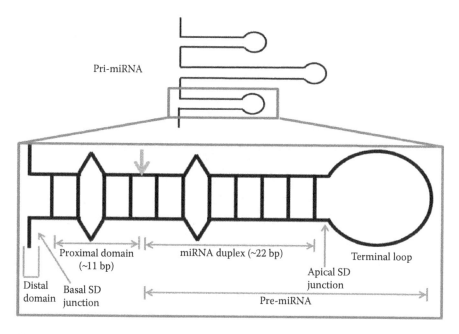

FIGURE 1.2 Structure of pri-miRNA. A pri-miRNA may contain several stem-loop pre-miRNAs and the flanking region. Arrow indicates the Drosha cleavage site.

the cleavage site (~11 bp from the basal single-stranded to double-stranded junction) on the pri-miRNA (Han et al., 2006). But now it has been established that both Drosha and DGCR8 play a role in the recognition and proper alignment of pri-miRNA (Nguyen et al., 2015). A pri-miRNA is recognized as a substrate by the microprocessor only if it fulfills certain requirements including the presence of a ~33 bp stem, a ~10 bp terminal loop, single-stranded-to-double-stranded RNA (SD) junction in the basal and apical regions, a single-stranded stretch in the distal domain of stem, a double-stranded stretch in the proximal domain of stem and a miRNA duplex (Feng et al., 2011; Han et al., 2006; Ma et al., 2013; Zeng and Cullen, 2005; Zeng et al., 2005; Zhang and Zeng, 2010).

The binding of microprocessor to pri-miRNA occurs as a heterotrimer, consisting of a Drosha and two DGCR8 subunits, and depends mostly on the structure of pri-miRNA, and to some extent, its sequence (Nguyen et al., 2015). Drosha consists of N-terminal domain, two RNase III domains, the central domain, and one double-stranded RNA-binding domain (dsRBD) (Han et al., 2004; Lee et al., 2003). The protein can identify a 'UG' motif near the SD junction at the base of the stem or at the terminal loop and by itself, in the absence of DGCR8, can cleave around 11 bp away from this junction, causing non-productive processing of the pri-miRNA (Auyeung et al., 2013; Han et al., 2004, 2006; Lee et al., 2003; Nguyen et al., 2015).

DGCR8 helps in the productive processing of pri-miRNAs by binding to the terminal loop and directing Drosha towards the base of pri-miRNA (Nguyen et al., 2015). DGCR8 contains many domains for exerting its function, including the nuclear localization signal (NLS), an RNA-binding heme domain (Rhed) in the

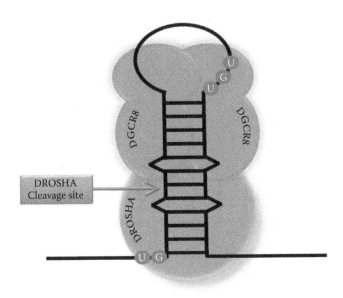

FIGURE 1.3 Binding of microprocessor complex to pri-miRNA. DROSHA recognizes a "UG" motif in the basal region and binds to the lower stem. DGCR8 forms a dimer and recognizes a "UGU" motif in the terminal loop and binds to the loop and stem regions.

center, two dsRBDs, and the C-terminal tail region (Yeom et al., 2006). It recognizes and binds to a "UGU" motif in the terminal loop via Rhed and to the upper stem region via dsRBDs that gives an asymmetry to the pri-miRNA structure and provides directionality for the binding of Drosha at the basal SD region. And at the same time, DGCR8 binds to and stabilizes Drosha via the C-terminal tail (Auyeung et al., 2013; Faller et al., 2010; Quick-Cleveland et al., 2014; Yeom et al., 2006). Thus, the coordination between Drosha and DGCR8 helps to recognize the suitable substrate and attain productive processing of pri-miRNAs to give rise to the correct pre-miRNA (see Figure 1.3).

1.4 CONTROL OF miRNA BIOGENESIS

1.4.1 AT THE NUCLEAR PROCESSING STEP

miRNAome of a cell is different for different cell/tissue types. Further, it can be fine-tuned according to the developmental stage or cellular condition (see Figure 1.4). The core microprocessor complex itself processes different pri-miRNAs with different efficiency. For example, due to the absence of a "UGU" motif in the apical region of pri-miR-16-1, its processing largely depends on Drosha (Han et al., 2006).

Several proteins alter the processing of specific pri-miRNA transcripts. For example, nucleolin protein is specifically required for the processing of pri-miR-15a/16 transcript (Pickering et al., 2011). Nuclear nucleolin interacts with Drosha and DGCR8 and also binds to pri-miR-15a/16 and enables its processing by the microprocessor complex (Pickering et al., 2011).

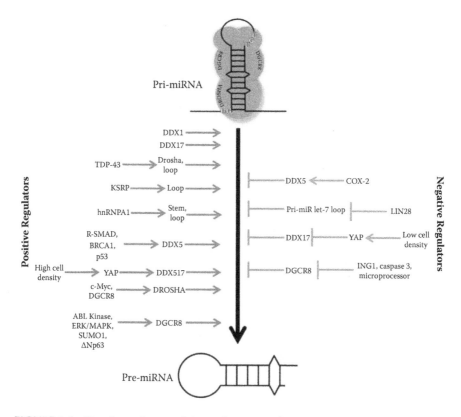

FIGURE 1.4 Regulatory factors of the nuclear processing step.

As stated earlier, the microprocessor complex consists primarily of Drosha and DGCR8 proteins. However, some accessory proteins, including DEAD box RNA helicases, DEAH box RNA helicases, double-stranded RNA binding proteins, hnRNP proteins and Ewing's sarcoma family of proteins containing an RNA recognition motif (RRM), are also part of the microprocessor complex (Gregory et al., 2004). These accessory proteins can further modulate the pri-miRNA processing by altering the affinity of the microprocessor for specific miRNAs.

DDX5 can both promote and inhibit processing of groups of pri-miRNAs under the influence of various factors. For example, the cyclooxygenase 2 (COX-2) protein interacts with the microprocessor complex through DDX5 and suppresses the processing of pri-miR-183, thus increasing the levels of insulin receptor substrate, IRS-1, resulting in protection against insulin resistance (Motiño et al., 2015). Furthermore, BRCA1 notably binds to DDX5, Drosha and the basal region of a subset of pri-miRNAs and increases their processing (Kawai and Amano, 2012).

The terminal loop of pri-miRNAs is the binding site for many microprocessor cofactors and is a means to modulate the processing. In a study it was seen that the DEAD-box RNA helicase DDX1 could bind to the terminal loop of a subset of pri-miRNAs and escort them to the microprocessor complex by its Drosha binding ability, thus boosting their processing during neocarzinostatin induced DNA damage (Chakravarti et al., 2014).

It is seen that many times the level of individual miRNAs of the same cluster that are transcribed as a single primary transcript are not similar. Different levels of premature miRNAs of a cluster are reflected by their different processing patterns. Proteins binding to a specific structure or sequence of pri-miRNAs might be the crucial factor for the uneven processing of clustered miRNAs (Remenyi et al., 2016).

DEAD-box RNA helicases also help in the differential processing of clustered miRNAs. DDX17 preferentially binds to the 18 base, relatively closed terminal loop of pri-miR-132 and favors its processing over its co-transcribed miRNA pri-miR-212 (Remenyi et al., 2016).

Other than the microprocessor cofactors, many regulatory proteins bind to their pri-miRNA targets via the loop region. One such example is that of TAR DNA-binding protein-43 (TDP-43) that binds to the Drosha complex and to the loop of specific pri-miRNA, thereby stabilizing their interaction and promoting the biogenesis of neuronal miRNAs (Kawahara and Mieda-Sato, 2012). Similarly, KH-type splicing regulatory protein (KSRP), after being phosphorylated by ATM, can stabilize the pri-miRNA/ microprocessor interaction and increase pri-miRNA processing (Trabucchi et al., 2009).

The terminal loop not only helps in recruiting pri-miRNA to the microprocessor, but also plays a role in sequestering the pri-miRNA to prevent its processing. For example, LIN28 uses this method to regulate pri-miRlet-7 processing (Newman et al., 2008). LIN28 promotes self-renewal of stem cells and let-7 promotes cell differentiation (Büssing et al., 2008; Viswanathan and Daley, 2010). LIN28B is known to sequester pri-miR let-7 in the nucleoli during early stages of development and pri-miR let-7 processing only begins with differentiation due to the reduction of LIN28B levels (Piskounova et al., 2011).

hnRNP A1 is another accessory protein of the microprocessor complex that binds exclusively to the terminal loop and stem regions of pri-miRNA-18a of the *mir-17* cluster and enhances its processing by the Drosha-DGCR8 complex (Guil and Cáceres, 2007; Michlewski et al., 2010). hnRNP A1 recognizes its consensus sequence UAGGGA/U present in the stem and loop of pri-miR-18a of the *miR-17* cluster, causes relaxation of the Drosha cleavage site, and thereby facilitates the processing of only pri-miR-18a (Michlewski et al., 2010).

miRNA biogenesis is regulated at post-transcriptional level by signaling cascades as well. Under the influence of TGF-β and BMP (bone morphogenetic protein) mediated pathways, R-SMADs bind to the stem region of pri-miR-21 and also interact with p68, thus, recruiting the pri-miRNA to microprocessor complex promoting its processing (Davis et al., 2008, 2010; Kang et al., 2012).

In a similar way, the p53 pathway also regulates miRNA biogenesis by binding to the p68 and to a subset of pri-miRNAs (pri-miR-16-1, pri-miR-143 and pri-miR-145), which enhances their processing by the Drosha complex (Suzuki et al., 2009).

The levels of chief microprocessor components, Drosha and DGCR8, are also subjected to alteration by various factors. Other than activation of miRNA genes, c-Myc also transactivates Drosha expression by binding to the E box present in *DROSHA* promoter, thus, increasing overall miRNA processing (Wang et al., 2013). During the differentiation of the epidermal cells, the N-terminal truncated form of p63 known as ΔNp63 causes transcriptional activation of *DGCR8* which leads to the formation of a whole new set of differentiation-promoting miRNAome (Chakravarti et al., 2014). Drosha and DGCR8 are themselves known to regulate each other post-transcriptionally to maintain the equilibrium of the microprocessor (Han et al., 2009). DGCR8 is known to stabilize the structure of Drosha protein (Han et al., 2009). The microprocessor complex recognizes a stem-loop hairpin in the 5'UTR of DGCR8 mRNA and causes its cleavage, thus, depleting the levels of DGCR8 (Han et al., 2009; Triboulet et al., 2009). Further, caspase 3 proteolytically cleaves DGCR8 at the heme-binding domain, leading to decreased expression of miRNAs in apoptotic cells (Gong et al., 2012).

The tumor suppressor Hippo signaling pathway modulates miRNA processing via YAP in a cell-density dependent manner (Mori et al., 2014). At low cell density, YAP is translocated inside the nucleus where it sequesters DDX17 and abrogates pri-miRNA processing. At high cell density however, YAP is retained in the cytoplasm, promoting the association of DDX17 with the microprocessor complex and facilitating pri-miRNA processing (Mori et al., 2014).

Post-translational modifications also affect the stability as well as activity of the microprocessor complex. SUMO1 is known to cause SUMOylation of DGCR8 at K^{707} that prevents its degradation by the proteasomal degradation pathway (Zhu et al., 2015). DGCR8 can also be stabilized by the ERK/MAPK pathway that causes phosphorylation of DGCR8 at multiple sites (Herbert et al., 2013). HDAC1 alters the activity of the microprocessor by deacetylation of lysine residues in the RNA binding domains of DGCR8, thus, increasing its affinity towards pri-miRNA transcripts (Wada et al., 2012). The tumor suppressor ING1 has been shown to recruit HDACs to DGCR8 promoter causing deacetylation of H3 and H4 that finally leads to down-regulation of both DGCR8 and a subset of miRNAs (Gómez-Cabello et al., 2010). The ABL kinase phosphorylates DGCR8 for specific processing of pri-miR-34c (Tu et al., 2015). The positioning of Drosha-DGCR8 on pri-miRNA is crucial for its processing. The flanking sequences at the base of stem of pri-miR-34c preferentially bind to DGCR8 instead of Drosha, leading to the formation of stable DGCR8-RNA complex, which is resistant towards Drosha processing (Tu et al., 2015). ABL kinase phosphorylates DGCR8 at Tyr^{267} because of which DGCR8 is unable to crosslink to the flanking pri-miR-34c sequence, leaving it accessible for Drosha, thus promoting its processing (Tu et al., 2015).

Adenosine deaminases acting on RNA (ADARs) are a group of proteins known to alter the miRNA levels by modulating the stability of miRNA biogenesis pathway intermediates (Yang et al., 2006). These proteins convert the adenosine residues in miRNA to inosine residues, leading to the degradation of the miRNA, thus maintaining the miRNA levels in the cell (Yang et al., 2006).

1.4.2 Nuclear Export of Pre-miRNA

After processing of the pri-miRNA by the microprocessor, the resulting pre-miRNA is exported out of the nucleus. The export of pre-miRNA is carried out by a complex of exportin5 (Exp5) and Ran-GTP protein (see Figure 1.5) (Bohnsack et al., 2004; Lund et al., 2004; Yi et al., 2003). Exp5 recognizes a short 3′ overhang in the pre-miRNA and binds to the stem region (Yi et al., 2003). The presence of a 3′

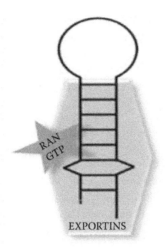

FIGURE 1.5   XPO5: Ran-GTP Complex with pre-miRNA.

overhang and >16 bp stem region is mandatory for recognition of a stem-loop RNA as a potential cargo by Exp5 (Zeng and Cullen, 2004). Exp5 binds to the stem region via its baseball mitt-like structure and to the 3′ overhang via its tunnel like structure (Okada et al., 2009).

Exp5 belongs to karyopherin family of transport proteins (Brownawell and Macara, 2002). The karyopherin family members are a family of nuclear exporters that bind to their "cargo" and the cofactor Ran-GTP in the nucleus. The Ran-GTP is subsequently hydrolyzed in the cytoplasm, inducing the release of the cargo (Bohnsack et al., 2004). Exp5 transports many RNAs out of the nucleus, including adenovirus VA1 RNA, human Y1 RNA,tRNA and the pre-miRNA (Calado et al., 2002; Gwizdek et al., 2003; Yi et al., 2003).

It was shown that downregulation of Exp5 reduces mature miRNA levels without accumulation of pre-miRNA, suggesting a role of Exp5 in stabilizing the pre-miRNA (Yi et al., 2003; Zeng and Cullen, 2004).

Nuclear export of pre-miRNAs is also affected by many regulatory factors. Recently, it was shown that during neocarzinostatin (NCS) induced DNA damage, nuclear export of pre-miRNAs increased in an ATM-dependent manner (Wan et al., 2013). ATM activated Akt phosphorylates nuclear pore protein NUP153 enhancing its interaction with Exp5, thus leading to improved pre-miRNA export (Wan et al., 2013).

1.5   PROCESSING OF THE PRECURSOR miRNA TRANSCRIPT

After the nuclear export of pre-miRNA, it is further cleaved near the terminal loop by Dicer to form the ~22 base pair miRNA duplex (Bernstein et al., 2001). Thus, one end of mature miRNA is decided by Drosha complex and the other end is decided by Dicer.

The domains of Dicer protein include an N-terminal helicase domain, a domain of unknown function (DUF283), a Piwi-Argonaute-Zwille (PAZ) domain, a C-terminal dsRNA binding domain and two RNase III domains (Macrae et al., 2006; Zhang et al., 2004). Sequentially, the PAZ domain first recognizes and binds to the 5′ and 3′ end of pre-miRNA. This is followed by the interaction of the N-terminal helicase domain with the terminal loop of pre-miRNA (Park et al., 2011; Tsutsumi et al., 2011). The helicase domain checks the terminal loop size and the PAZ domain measures the distance from the 3′ end to the terminal loop. Together, they help to align the pre-miRNA in the catalytic site of Dicer and also to distinguish between miRNA and siRNA substrates (Ma et al., 2004; Park et al., 2011; Tsutsumi et al., 2011). Dicer binds to the 5′ end only if it is thermodynamically unstable and then cleaves 22 base away from the 5′ end of the pre-miRNA (Park et al., 2011). After the proper alignment of pre-miRNA, the RNase III domains form an intramolecular dimer which acts as the catalytic center to cleave the pre-miRNA to generate miRNA duplex with 2-nucleotide(nt) 3′ overhangs (Zhang et al., 2004).

The accessory proteins of mammalian Dicer include TRBP (Transactivation response element RNA Binding Protein) and PACT (Protein ACTivator of PKR) (Haase et al., 2005; Lee et al., 2006). Both TRBP and PACT help in pre-miRNA cleavage and RISC assembly (see Figure 1.6) (Chendrimada et al., 2005; Lee et al., 2013, 2006).

TRBP is a double-stranded RNA binding protein containing three dsRBDs of which two are situated at the N-terminus and one is present at the C-terminal domain of the protein. TRBP binds to RNA via its two N-terminal dsRBDs while it uses its C-terminal dsRBD to interact with other proteins including Dicer and PACT in an RNA-independent manner (Daniels et al., 2009; Daviet et al., 2000; Laraki et al., 2008). TRBP tunes the length of mature miRNAs as depletion of TRBP leads to generation of isomiRs truncated at the 5′ end by 1 nt and with shifted seed sequence

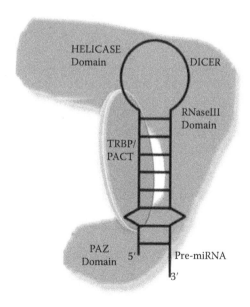

FIGURE 1.6 Binding of DICER/TRBP/PACT Complex to pre-miRNAs.

(Fukunaga et al., 2012; Kim et al., 2014). Thus, TRBP seems to be an important cofactor of the Dicer ribonuclease. Surprisingly, it was shown that TRBP is required for the accurate processing of only a subset of miRNAs (Kim et al., 2014).

TRBP and PACT have specific non-redundant roles in determining the cleavage sites to produce miRNAs with varying length (isomiRs) and to distinguish between pre-siRNA and pre-miRNA substrates for the Dicer enzyme (Lee et al., 2013). It was reported that PACT inhibits pre-siRNA processing more strongly than TRBP (Lee et al., 2013).

1.5.1 CONTROL OF miRNA BIOGENESIS AT THE CYTOPLASMIC PROCESSING STEP

Dicer activity can be regulated either by directly affecting Dicer or by modulation of one of its binding partners. Dicer levels are maintained at homeostasis by binding of let-7 miRNA that causes degradation of Dicer mRNA by a negative feedback loop (Tokumaru et al., 2008). The Dicer cofactor TRBP can modify Dicer activity and cause selective processing of specific pre-miRNAs. Further, it is also involved in the generation of isomiRs (Fukunaga et al., 2012; Lee and Doudna, 2012).

TRBP can be influenced by various modifiers of Dicer activity. MAPK/ERK causes phosphorylation of TRBP that leads to preferential upregulation of growth promoting miRNAs (Paroo et al., 2009). Another TRBP modulator, Angiopoietin-1, triggers the phosphorylation of TRBP by S6K2 kinase, causing its activation that results in enhanced expression of several miRNAs in primary human dermal lymphatic endothelial cells (HDLEC) (Warner et al., 2016).

Many proteins that modulate the Drosha processing step are also involved in the regulation of pre-miRNA processing by Dicer. For example, KSRP binds to the terminal loop in pri-miRNA to modulate Drosha processing as well as also binds to the terminal loop of specific pre-miRNAs in cytoplasm and enhances their processing by Dicer (Trabucchi et al., 2009; Zhang et al., 2011). LIN28B protein binds to pri-miRNA in the nucleus and LIN28A protein binds to the terminal loop of pre-let-7 in cytoplasm and abrogates its processing by Dicer (Piskounova et al., 2011). Just like the N-terminal truncated form of p63 (ΔNp63) transcriptionally activates DGCR8, the full length p63 (Tap63) can also bind to and transcriptionally activate the promoters of DICER and miR-130b, thereby suppressing tumorigenesis and metastasis in MEF cells (Su et al., 2010). Similarly, TDP-43, besides regulating the Drosha processing step, also interacts with Dicer complex in the cytoplasm and terminal loop of pre-miRNAs to promote their processing into mature miRNA duplexes (Kawahara and Mieda-Sato, 2012).

Dicer protein is subjected to auto-inhibition via its internal helicase domain (Ma et al., 2008). It was seen that deletion of the helicase domain causes activation of Dicer suggesting that the helicase domain might inhibit the catalytic activity of Dicer until some factor induces structural rearrangement of Dicer, thus causing its activation (Ma et al., 2008).

Dicer mRNA levels can be optimized by RISC complex guided by let-7 miRNA where let-7 causes downregulation of Dicer mRNA and protein levels by a negative feedback loop ultimately altering the level of other miRNAs (Tokumaru et al., 2008).

The levels of pre-miRNA can also be controlled by its degradation via ribonuclease MCPIP1, that binds to the terminal loop of pre-miRNAs and cleaves them

(Suzuki et al., 2011). Thus, MCPIP1 counteracts Dicer to maintain the level of pre-miRNAs (Suzuki et al., 2011).

1.5.2 RNA INDUCED SILENCING COMPLEX

After the Dicer processing step, the miRNA duplex is loaded onto the Argonaute (Ago2) protein to form the effector complex, RISC (RNA induced silencing complex). Argonaute (AGO) proteins are the primary component of the RISC complex and are the effectors of miRNA-mediated gene silencing (see Figure 1.7).

Argonaute proteins are divided into two subclasses based on sequence homology: i) the Ago subfamily resembles Arabidopsis AGO1 protein and ii) the Piwi subfamily resembles Drosophila PIWI protein (Carmell et al., 2002). In humans, the Ago subfamily contains AGO1, AGO2, AGO3 and AGO4, out of which only AGO2 is able to exert RNA catalytic activity. However, all the AGO proteins are capable of inducing translational repression (Meister, 2013). The expression of PIWI subfamily proteins is mainly restricted to the germline (Carmell et al., 2002).

AGO proteins are characterized by the presence of an N-PAZ lobe, consisting of the N-terminal helicase domain and the PAZ domain and the MID-PIWI lobe, containing a MID (middle) domain and a PIWI domain (Jinek and Doudna, 2009). The two domains are connected by a hinge that undergoes structural rearrangement as a result of binding of miRNA duplex (Jinek and Doudna, 2009).

The Dicer processing step and formation of RISC complex is a coupled phenomenon. At first, AGO2 associates with Dicer, pre-miRNA and TRBP/PACT to form the RISC Loading Complex (RLC) which allows the tight coupling of Dicer cleavage

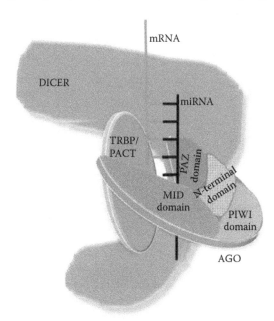

FIGURE 1.7 The RNA-induced Silencing Complex.

to the incorporation of miRNA into the RISC complex (Chendrimada et al., 2005; Lee et al., 2006). After Dicer cleavage, the miRNA duplex is transferred from Dicer to AGO2, a step that is achieved by the chaperone HSP90 in humans (Iwasaki et al., 2010). HSP90 uses the energy from ATP hydrolysis to keep AGO2 in an open conformation so as to accommodate the miRNA duplex properly (Johnston et al., 2010). The MID domain of AGO2 binds to the 5′ end and its PAZ domain binds to the 3′ end of the duplex; subsequently, the N-terminal helicase domain causes unwinding of the miRNA duplex (Jinek and Doudna, 2009; Meister, 2013). Afterwards, the more thermodynamically stable strand (passenger strand or miRNA*) is cleaved by PIWI domain, and the less thermodynamically stable strand (guide strand or miRNA) remains associated with AGO2 to form the RISC (Khvorova et al., 2003; Noland and Doudna, 2013; Schwarz et al., 2003). The cleaved passenger strand is removed from the RLC by component 3 promoter of RISC (C3PO) that is also known as translin (Liu et al., 2009). The miRNA then guides the RISC complex to the 3′UTR of its target mRNA, where it binds with partial complementarity and leads to translational repression or mRNA degradation (Huntzinger and Izaurralde, 2011). The sequence of miRNA, with which it binds to its target, is called the seed sequence and is composed of nucleotides 2 to 7 in the 5′ end of the miRNA.

1.5.3 REPRESSION/DECAY OF TARGET MRNA

The translational repression/decay of target mRNA is accomplished by the RISC complex with the help of GW182 protein (see Figure 1.8) (Liu et al., 2005). The domains and motifs of GW182 includes an N-terminal domain containing glycine-tryptophan (GW) repeats, a C-terminal domain with GW repeats, a poly(A)-binding

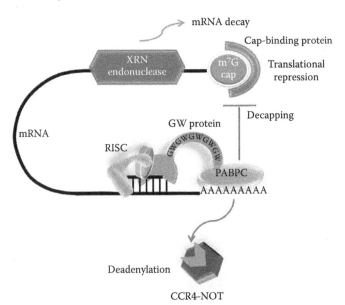

FIGURE 1.8 Translational repression and decay of target mRNA by the RISC assembly.

protein-interacting motif 2 (PAM2) motif, and an RNA recognition motif (RRM) (Eulalio et al., 2009). AGO2 binds via its PIWI domain to the N-terminal domain of GW protein (Huntzinger and Izaurralde, 2011). Consequently, GW protein uses its PAM2 and RRM motifs to interact with a poly(A)-binding protein C (PABPC) on the poly(A) tail of mRNA (Zekri et al., 2009). PABPC is known to interact with the cap-binding complex that leads to circularization of mRNA promoting translation. When GW protein binds to PABPC, it causes repression of translation by inhibiting circularization of the target mRNA (Zekri et al., 2009). Next, the C-terminal GW motif interacts with and recruits the deadenylase CCR4–NOT complex to the mRNA which induces excision of the poly(A) tail (Chekulaeva et al., 2011). This leads to destabilization of the mRNA and the destabilized mRNA undergoes decapping and degradation by XRN1 endonuclease (Huntzinger and Izaurralde, 2011).

1.6 CONCLUSION

Each cell has its own specific miRNAome wherein the small ~22nt miRNA molecules modulate gene expression either by promoting mRNA degradation or by blocking translation. miRNA biogenesis occurs both via canonical as well as noncanonical pathways. Both the pathways are tightly regulated both at transcriptional as well as post-transcriptional steps. Further, regulation via feedback as well as feed forward pathways have also been described. The regulation of the miRNA biogenesis is critical as dysregulation has been linked to diseases ranging from cancer to neurodegenerative disorders. The role of miRNAome and how their dysregulation results in diseases will be discussed in the ensuing chapters.

REFERENCES

Altuvia, Y., Landgraf, P., Lithwick, G. et al. 2005. Clustering and conservation patterns of human microRNAs. *Nucleic Acids Res.* *33*: 2697–2706.

Ardekani, A.M., and Naeini, M.M. 2010. The role of microRNAs in human diseases. Avicenna *J. Med. Biotechnol. 2*: 161–179.

Auyeung, V.C., Ulitsky, I., McGeary, S.E. et al. 2013. Beyond secondary structure: Primary-sequence determinants license pri-miRNA hairpins for processing. *Cell 152*: 844–858.

Bartel, D.P. 2004. MicroRNAs: Genomics, biogenesis, mechanism, and function. *Cell 116*: 281–297.

Baskerville, S., and Bartel, D.P. 2005. Microarray profiling of microRNAs reveals frequent coexpression with neighboring miRNAs and host genes. *RNA N. Y. N 11*: 241–247.

Bernstein, E., Caudy, A.A., Hammond, S.M. et al. 2001. Role for a bidentateribonuclease in the initiation step of RNA interference. *Nature 409*: 363–366.

Bohnsack, M.T., Czaplinski, K., and Gorlich, D. 2004. Exportin 5 is a RanGTP-dependent dsRNA-binding protein that mediates nuclear export of pre-miRNAs. *RNA N. Y. N 10*: 185–191.

Borchert, G.M., Lanier, W., and Davidson, B.L. 2006. RNA polymerase III transcribes human microRNAs. *Nat. Struct. Mol. Biol. 13*: 1097–1101.

Brownawell, A.M., and Macara, I.G. 2002. Exportin-5, a novel karyopherin, mediates nuclear export of double-stranded RNA binding proteins. *J. Cell Biol. 156*: 53–64.

Büssing, I., Slack, F.J., and Grosshans, H. 2008. let-7 microRNAs in development, stem cells and cancer. *Trends Mol. Med. 14*: 400–409.

Calado, A., Treichel, N., Müller, E.C. et al. 2002. Exportin-5-mediated nuclear export of eukaryotic elongation factor 1A and tRNA. *EMBO J.* **21**: 6216–6224.

Carmell, M.A., Xuan, Z., Zhang, M.Q. et al. 2002. The Argonaute family: Tentacles that reach into RNAi, developmental control, stem cell maintenance, and tumorigenesis. *Genes Dev.* **16**: 2733–2742.

Chakravarti, D., Su, X., Cho, M.S. et al. 2014. Induced multipotency in adult keratinocytes through down-regulation of ΔNp63 or DGCR8. *Proc. Natl. Acad. Sci. U. S. A.* **111**: E572–E581.

Chang, T.-C., Wentzel, E.A., Kent, O.A. et al. 2007. Transactivation of miR-34a by p53 broadly influences gene expression and promotes apoptosis. *Mol. Cell* **26**: 745–752.

Chang, T.-C., Yu, D., Lee, Y.-S. et al. 2008. Widespread microRNA repression by Myc contributes to tumorigenesis. *Nat. Genet.* **40**: 43–50.

Chekulaeva, M., Mathys, H., Zipprich, J.T. et al. 2011. miRNA repression involves GW182-mediated recruitment of CCR4-NOT through conserved W-containing motifs. *Nat. Struct. Mol. Biol* **18**: 1218–1226.

Chendrimada, T.P., Gregory, R.I., Kumaraswamy, E. et al. 2005. TRBP recruits the Dicer complex to Ago2 for microRNA processing and gene silencing. *Nature* **436**: 740–744.

Corney, D.C., Flesken-Nikitin, A., Godwin, A.K. et al. 2007. MicroRNA-34b and MicroRNA-34c are targets of p53 and cooperate in control of cell proliferation and adhesion-independent growth. *Cancer Res.* **67**: 8433–8438.

Daniels, S.M., Melendez-Peña, C.E., Scarborough, R.J. et al. 2009. Characterization of the TRBP domain required for Dicer interaction and function in RNA interference. *BMC Mol. Biol.* **10**: 38.

Daugaard, I., and Hansen, T.B. 2017. Biogenesis and function of Ago-associated RNAs. *Trends Genet.* **33**: 208–219.

Daviet, L., Erard, M., Dorin, D. et al. 2000. Analysis of a binding difference between the two dsRNA-binding domains in TRBP reveals the modular function of a KR-helix motif. *Eur. J. Biochem.* **267**: 2419–2431.

Davis, B.N., Hilyard, A.C., Lagna, G. et al. 2008. SMAD proteins control DROSHA-mediated microRNA maturation. *Nature* **454**: 56–61.

Davis, B.N., Hilyard, A.C., Nguyen, P.H. 2010. Smad proteins bind a conserved RNA sequence to promote microRNA maturation by Drosha. *Mol. Cell* **39**: 373–384.

Denli, A.M., Tops, B.B.J., Plasterk, R.H.A. et al. 2004. Processing of primary microRNAs by the Microprocessor complex. *Nature* **432**: 231–235.

Eulalio, A., Tritschler, F., and Izaurralde, E. 2009. The GW182 protein family in animal cells: New insights into domains required for miRNA-mediated gene silencing. *RNA* **15**: 1433–1442.

Faller, M., Toso, D., Matsunaga, M. et al. 2010. DGCR8 recognizes primary transcripts of microRNAs through highly cooperative binding and formation of higher-order structures. *RNA* **16**: 1570–1583.

Feng, Y., Zhang, X., Song, Q. et al. 2011. Drosha processing controls the specificity and efficiency of global microRNA expression. *Biochim. Biophys. Acta* **1809**: 700–707.

Fukunaga, R., Han, B.W., Hung, J.-H. et al. 2012. Dicer partner proteins tune the length of mature miRNAs in flies and mammals. *Cell* **151**: 533–546.

Glozak, M.A., and Seto, E. 2007. Histone deacetylases and cancer. *Oncogene* **26**: 5420–5432.

Gómez-Cabello, D., Callejas, S., Benguría, A. et al. 2010. Regulation of the microRNA processor DGCR8 by the tumor suppressor ING1. *Cancer Res.* **70**: 1866–1874.

Gong, M., Chen, Y., Senturia, R. et al. 2012. Caspases cleave and inhibit the microRNA processing protein DiGeorge Critical Region 8. *Protein Sci. Publ. Protein Soc.* **21**: 797–808.

Gray, S.G., and Teh, B.T. 2001. Histone acetylation/deacetylation and cancer: An "open" and "shut" case? *Curr. Mol. Med.* **1**: 401–429.

Gregory, R.I., Yan, K., Amuthan, G. et al. 2004. The microprocessor complex mediates the genesis of microRNAs. *Nature* **432**: 235–240.

Gregory, R.I., Chendrimada, T.P., Cooch, N. et al. 2005. Human RISC couples microRNA biogenesis and posttranscriptional gene silencing. *Cell* **123**: 631–640.

Guil, S., and Cáceres, J.F. 2007. The multifunctional RNA-binding protein hnRNP A1 is required for processing of miR-18a. *Nat. Struct. Mol. Biol.* **14**: 591–596.

Gwizdek, C., Ossareh-Nazari, B., Brownawell 2003. Exportin-5 mediates nuclear export of minihelix-containing RNAs. *J. Biol. Chem.* **278**: 5505–5508.

Haase, A.D., Jaskiewicz, L., Zhang, H. et al. 2005. TRBP, a regulator of cellular PKR and HIV-1 virus expression, interacts with Dicer and functions in RNA silencing. *EMBO Rep.* **6**: 961–967.

Han, J., Lee, Y., Yeom, K.-H. et al. 2004. The Drosha-DGCR8 complex in primary microRNA processing. *Genes Dev.* **18**: 3016–3027.

Han, J., Lee, Y., Yeom, K.-H. et al. 2006. Molecular basis for the recognition of primary microRNAs by the Drosha–DGCR8 complex. *Cell* **125**: 887–901.

Han, J., Pedersen, J.S., Kwon, S.C. et al. 2009.Posttranscriptional cross regulation between Drosha and DGCR8. *Cell* **136**: 75–84.

Han, C., Liu, Y., Wan, G. 2014. The RNA-binding protein DDX1 promotes primary microRNA maturation and inhibits ovarian tumor progression. *Cell Rep.* **8**: 1447–1460.

Herbert, K.M., Pimienta, G., DeGregorio, S.J. et al. 2013. Phosphorylation of DGCR8 increases its intracellular stability and induces a progrowth miRNA profile. *Cell Rep.* **5**: 1070–1081.

Huntzinger, E., and Izaurralde, E. 2011. Gene silencing by microRNAs: Contributions of translational repression and mRNA decay. *Nat. Rev. Genet.* **12**: 99–110.

Hutvágner, G., McLachlan, J., Pasquinelli, A.E. et al. 2001. A cellular function for the RNA-interference enzyme dicer in the maturation of the let-7 small temporal RNA. *Science* **293**: 834–838.

Iwasaki, S., Kobayashi, M., Yoda, M. et al. 2010. Hsc70/Hsp90 chaperone machinery mediates ATP-dependent RISC loading of small RNA duplexes. *Mol. Cell* **39**: 292–299.

Jinek, M., and Doudna, J.A. 2009. A three-dimensional view of the molecular machinery of RNA interference. *Nature* **457**: 405–412.

Johnston, M., Geoffroy, M.-C., Sobala, A. et al. 2010. HSP90 protein stabilizes unloaded argonaute complexes and microscopic P-bodies in human cells. *Mol. Biol. Cell* **21**: 1462–1469.

Kang, H., Davis-Dusenbery, B.N., Nguyen, P.H. et al. 2012. Bone morphogenetic protein 4 promotes vascular smooth muscle contractility by activating microRNA-21 (miR-21), which down-regulates expression of family of dedicator of cytokinesis (DOCK) proteins. *J. Biol. Chem.* **287**, 3976–3986.

Kato, M., Putta, S., Wang, M. 2009. TGF-β activates Akt kinase via a microRNA-dependent amplifying circuit targeting PTEN. *Nat. Cell Biol.* **11**: 881–889.

Kawahara, Y., and Mieda-Sato, A. 2012. TDP-43 promotes microRNA biogenesis as a component of the Drosha and Dicer complexes. *Proc. Natl. Acad. Sci. U. S. A.* **109**: 3347–3352.

Kawai, S., and Amano, A. 2012. BRCA1 regulates microRNA biogenesis via the DROSHA microprocessor complex. *J. Cell Biol.* **197**: 201–208.

Kent, O.A., Chivukula, R.R., Mullendore, M. et al. 2010. Repression of the miR-143/145 cluster by oncogenic Ras initiates a tumor-promoting feed-forward pathway. *Genes Dev.* **24**: 2754–2759.

Ketting, R.F., Fischer, S.E.J., Bernstein, E. et al. 2001. Dicer functions in RNA interference and in synthesis of small RNA involved in developmental timing in C. elegans. *Genes Dev.* **15**: 2654–2659.

Khvorova, A., Reynolds, A., and Jayasena, S.D. 2003. Functional siRNAs and miRNAs exhibit strand bias. *Cell* **115**: 209–216.

Kim, V.N., Han, J., and Siomi, M.C. 2009. Biogenesis of small RNAs in animals. *Nat. Rev. Mol. Cell Biol.* **10**: 126–139.

Kim, Y., Yeo, J., Lee, J.H. et al. 2014. Deletion of human tarbp2 reveals cellular microRNA targets and cell-cycle function of TRBP. *Cell Rep.* **9**: 1061–1074.

Landthaler, M., Yalcin, A., and Tuschl, T. 2004. The human DiGeorge syndrome critical region gene 8 and its *D. melanogaster* homolog are required for miRNA biogenesis. *Curr. Biol.* **14**: 2162–2167.

Laraki, G., Clerzius, G., Daher, A. et al. 2008. Interactions between the double-stranded RNA-binding proteins TRBP and PACT define the Medipal domain that mediates protein–protein interactions. *RNA Biol.* **5**: 92–103.

Lee, H.Y., and Doudna, J.A. 2012. TRBP alters human precursor microRNA processing in vitro. *RNA* **18**: 2012–2019.

Lee, H.Y., Zhou, K., Smith, A.M. et al. 2013. Differential roles of human Dicer-binding proteins TRBP and PACT in small RNA processing. *Nucleic Acids Res.* **41**: 6568–6576.

Lee, J., Li, Z., Brower-Sinning, R. 2007. Regulatory circuit of human microRNA biogenesis. *PLoSComput. Biol.* **3**: e67.

Lee, R.C., Feinbaum, R.L., and Ambros, V. 1993. The C. elegansheterochronic gene lin-4 encodes small RNAs with antisense complementarity to lin-14. *Cell* **75**: 843–854.

Lee, Y., Ahn, C., Han, J. et al. 2003. The nuclear RNase III Drosha initiates microRNA processing. *Nature* **425**: 415–419.

Lee, Y., Kim, M., Han, J. et al. 2004. MicroRNA genes are transcribed by RNA polymerase II. *EMBO J.* **23**: 4051–4060.

Lee, Y., Hur, I., Park et al. 2006. The role of PACT in the RNA silencing pathway. *EMBO J.* **25**: 522–532.

Liu, J., Rivas, F.V., Wohlschlegel, J. et al. 2005. A role for the P-body component, GW182, in microRNA function. *Nat. Cell Biol.* **7**: 1261–1266.

Liu, J., Zhang, C., Zhao, Y. 2017. MicroRNA control of p53. *J. Cell. Biochem.* **118**: 7–14.

Liu, Y., Ye, X., Jiang, F. et al. 2009. C3PO, an endoribonuclease that promotes RNAi by facilitating RISC activation. *Science* **325**: 750–753.

Lujambio, A., Calin, G.A., Villanueva, A. et al. 2008. A microRNA DNA methylation signature for human cancer metastasis. *Proc. Natl. Acad. Sci. U. S. A.* **105**: 13556–13561.

Lund, E., Güttinger, S., Calado, A. et al. 2004. Nuclear export of microRNA precursors. *Science* **303**: 95–98.

Ma, E., MacRae, I.J., Kirsch, J.F. et al. 2008. Autoinhibition of human dicer by its internal helicase domain. *J. Mol. Biol.* **380**: 237–243.

Ma, H., Wu, Y., Choi, J.-G. et al. 2013. Lower and upper stem-single-stranded RNA junctions together determine the Drosha cleavage site. *Proc. Natl. Acad. Sci. U. S. A.* **110**: 20687–20692.

Ma, J.-B., Ye, K., and Patel, D.J. 2004. Structural basis for overhang-specific small interfering RNA recognition by the PAZ domain. *Nature* **429**: 318–322.

Macrae, I.J., Li, F., Zhou, K. et al. 2006. Structure of Dicer and mechanistic implications for RNAi. *Cold Spring Harb. Symp. Quant. Biol.* **71**: 73–80.

Meister, G. 2013. Argonaute proteins: Functional insights and emerging roles. *Nat. Rev. Genet.* **14**: 447–459.

Michlewski, G., Guil, S., and Cáceres, J.F. 2010. Stimulation of pri-miR-18a processing by hnRNP A1. *Adv. Exp. Med. Biol.* **700**: 28–35.

Mori, M., Triboulet, R., Mohseni, M. et al. 2014. Hippo signaling regulates microprocessor and links cell-density-dependent miRNA biogenesis to cancer. *Cell* **156**: 893–906.

Motiño, O., Francés, D.E., Mayoral, R. et al. 2015. Regulation of microRNA 183 by cyclooxygenase 2 in liver is DEAD-box helicase p68 (DDX5) dependent: Role in insulin signaling. *Mol. Cell.Biol.* **35**: 2554–2567.

Newman, M.A., Thomson, J.M., and Hammond, S.M. 2008. Lin-28 interaction with the Let-7 precursor loop mediates regulated microRNA processing. *RNA 14*: 1539–1549.

Nguyen, T.A., Jo, M.H., Choi, Y.-G. et al. 2015. Functional anatomy of the human microprocessor. *Cell 161*: 1374–1387.

Noland, C.L., and Doudna, J.A. 2013. Multiple sensors ensure guide strand selection in human RNAi pathways. *RNA N. Y. N 19*: 639–648.

O'Donnell, K.A., Wentzel, E.A., Zeller, K.I. et al. 2005. c-Myc-regulated microRNAs modulate E2F1 expression. *Nature 435*: 839–843.

Okada, C., Yamashita, E., Lee, S.J. et al. 2009. A high-resolution structure of the pre-microRNA nuclear export machinery. *Science 326*: 1275–1279.

Park, J.-E., Heo, I., Tian, Y. et al. 2011. Dicer recognizes the 5′ end of RNA for efficient and accurate processing. *Nature 475*: 201–205.

Paroo, Z., Ye, X., Chen, S. 2009. Phosphorylation of the human microRNA-generating complex mediates MAPK/Erk signaling. *Cell 139*: 112–122.

Pickering, B.F., Yu, D., and Van Dyke, M.W. 2011. Nucleolin protein interacts with microprocessor complex to affect biogenesis of microRNAs 15a and 16. *J. Biol. Chem. 286*: 44095–44103.

Piskounova, E., Polytarchou, C., Thornton, J.E. et al. 2011. Lin28A and Lin28B inhibit let-7 microRNA biogenesis by distinct mechanisms. *Cell 147*: 1066–1079.

Quick-Cleveland, J., Jacob, J.P., Weitz, R. et al. 2014. The DGCR8 RNA-binding heme domain recognizes primary microRNAs by clamping the hairpin. *Cell Rep. 7*: 1994–2005.

Remenyi, J., Bajan, S., Fuller-Pace, F.V. et al. 2016. The loop structure and the RNA helicase p72/DDX17 influence the processing efficiency of the mice miR-132. *Sci. Rep. 6*: 22848. doi: 10.1038/srep22848.

Schwarz, D.S., Hutvágner, G., Du, T. et al. 2003. Asymmetry in the assembly of the RNAi enzyme complex. *Cell 115*: 199–208.

Spizzo, R., Nicoloso, M.S., Lupini, L. et al. 2010. miR-145 participates with TP53 in a death-promoting regulatory loop and targets estrogen receptor-alpha in human breast cancer cells. *Cell Death Differ. 17*: 246–254.

Su, X., Chakravarti, D., Cho, M.S. et al. 2010. TAp63 suppresses metastasis through coordinate regulation of Dicer and miRNAs. *Nature 467*: 986–990.

Suzuki, H.I., Yamagata, K., Sugimoto, K. et al. 2009. Modulation of microRNA processing by p53. *Nature 460*: 529–533.

Suzuki, H.I., Arase, M., Matsuyama, H. et al. 2011. MCPIP1 ribonuclease antagonizes Dicer and terminates microRNA biogenesis through precursor microRNA degradation. *Mol. Cell 44*: 424–436.

Sylvestre, Y., Guire, V.D., Querido, E. et al. 2007. An E2F/miR-20a autoregulatory feedback loop. *J. Biol. Chem. 282*: 2135–2143.

Talotta, F., Cimmino, A., Matarazzo, M.R. et al. 2009. An autoregulatory loop mediated by miR-21 and PDCD4 controls the AP-1 activity in RAS transformation. *Oncogene 28*: 73–84.

Tokumaru, S., Suzuki, M., Yamada, H. et al. 2008. let-7 regulates Dicer expression and constitutes a negative feedback loop. *Carcinogenesis 29*: 2073–2077.

Trabucchi, M., Briata, P., Garcia-Mayoral, M. et al. 2009. The RNA-binding protein KSRP promotes the biogenesis of a subset of microRNAs. *Nature 459*: 1010–1014.

Triboulet, R., Chang, H.-M., Lapierre, R.J. et al. 2009. Post-transcriptional control of DGCR8 expression by the microprocessor. *RNA N. Y. N 15*: 1005–1011.

Tsang, J., Zhu, J., and van Oudenaarden, A. 2007. MicroRNA-mediated feedback and feed-forward loops are recurrent network motifs in mammals. *Mol. Cell 26*: 753–767.

Tsutsumi, A., Kawamata, T., Izumi, N. et al. 2011. Recognition of the pre-miRNA structure by Drosophila Dicer-1. *Nat. Struct. Mol. Biol. 18*: 1153–1158.

Tu, C.-C., Zhong, Y., Nguyen, L. et al. 2015. The kinase ABL phosphorylates the microprocessor subunit DGCR8 to stimulate primary microRNA processing in response to DNA damage. *Sci. Signal. 8*: ra64.

Viswanathan, S.R., and Daley, G.Q. 2010. Lin28: A microRNA regulator with a macro role. *Cell* **140**: 445–449.

Wada, T., Kikuchi, J., and Furukawa, Y. 2012. Histone deacetylase 1 enhances microRNA processing via deacetylation of DGCR8. *EMBO Rep.* **13**: 142–149.

Wan, G., Zhang, X., Langley, R.R. et al. 2013. DNA-damage-induced nuclear export of precursor microRNAs is regulated by the ATM-AKT pathway. *Cell Rep.* **3**: 2100–2112.

Wang, X., Zhao, X., Gao, P. et al. 2013. c-Myc modulates microRNA processing via the transcriptional regulation of Drosha. *Sci. Rep.* **3**: 1942.

Warner, M.J., Bridge, K.S., Hewitson, J.P. et al. 2016. S6K2-mediated regulation of TRBP as a determinant of miRNA expression in human primary lymphatic endothelial cells. *Nucleic Acids Res.* **44**: 9942–9955.

Wilson, R.C., Tambe, A., Kidwell, M.A. et al. 2015. Dicer–TRBP complex formation ensures accurate mammalian microRNA biogenesis. *Mol. Cell* **57**: 397–407.

Winter, J., Jung, S., Keller, S. et al. 2009. Many roads to maturity: microRNA biogenesis pathways and their regulation. *Nat. Cell Biol.* **11**: 228–234.

Yamakuchi, M., Ferlito, M., and Lowenstein, C.J. 2008. miR-34a repression of SIRT1 regulates apoptosis. *Proc. Natl. Acad. Sci.* **105**: 13421–13426.

Yang, J.-S., and Lai, E.C. 2011. Alternative miRNA biogenesis pathways and the interpretation of core miRNA pathway mutants. *Mol. Cell* **43**: 892–903.

Yang, W., Chendrimada, T.P., Wang, Q. et al. 2006. Modulation of microRNA processing and expression through RNA editing by ADAR deaminases. *Nat. Struct. Mol. Biol.* **13**: 13–21.

Yeom, K.-H., Lee, Y., Han, J. et al. 2006. Characterization of DGCR8/Pasha, the essential cofactor for Drosha in primary miRNA processing. *Nucleic Acids Res.* **34**: 4622–4629.

Yi, R., Qin, Y., Macara, I.G. et al. 2003. Exportin-5 mediates the nuclear export of pre-microRNAs and short hairpin RNAs. *Genes Dev.* **17**: 3011–3016.

Zekri, L., Huntzinger, E., Heimstädt, S. et al. 2009. The silencing domain of GW182 interacts with PABPC1 to promote translational repression and degradation of microRNA targets and is required for target release. *Mol. Cell. Biol.* **29**: 6220–6231.

Zeng, Y., and Cullen, B.R. 2004. Structural requirements for pre-microRNA binding and nuclear export by Exportin 5. *Nucleic Acids Res.* **32**: 4776–4785.

Zeng, Y., and Cullen, B.R. 2005. Efficient processing of primary microRNA hairpins by Drosha requires flanking nonstructured RNA sequences. *J. Biol. Chem.* **280**: 27595–27603.

Zeng, Y., Yi, R., and Cullen, B.R. 2005. Recognition and cleavage of primary microRNA precursors by the nuclear processing enzyme Drosha. *EMBO J.* **24**: 138–148.

Zhang, X., and Zeng, Y. 2010. The terminal loop region controls microRNA processing by Drosha and Dicer. *Nucleic Acids Res.* **38**: 7689–7697.

Zhang, H., Kolb, F.A., Jaskiewicz, L. 2004. Single processing center models for human Dicer and bacterial RNase III. *Cell* **118**: 57–68.

Zhang, X., Wan, G., Berger, F.G. 2011. The ATM kinase induces microRNA biogenesis in the DNA damage response. *Mol. Cell* **41**: 371–383.

Zhu, C., Chen, C., Huang, J. 2015. SUMOylation at K707 of DGCR8 controls direct function of primary microRNA. *Nucleic Acids Res.* **43**: 7945–7960.

2 Computational Analysis of miRNAs, Their Target Sequences and Their Role in Gene Regulatory Networks

Pamchui Muiwo, Priyatama Pandey, and Alok Bhattacharya

CONTENTS

Introduction

Extensive studies have clearly shown the importance of microRNAs (miRNAs) in cellular and organismic biology of higher eukaryotes. miRNAs are a major member of a growing list of small non-regulatory RNAs that are produced through a processing event involving a number of different enzymes. These function by interacting with their cognate messenger RNAs (mRNAs) through formation of an RNA-protein complex known as RISC complex. Argonaut 2 (Ago2) is an important member of the RISC complex and is associated with miRNA:mRNA hybrid. miRNAs are distributed all over the genome, with the majority coming from intergenic regions, but many are also generated by a processing event from introns, noncoding RNAs and 3' UTR of a protein coding gene (Lagos-Quintana et al., 2001; Lee and Ambros, 2001). The major regulatory role of miRNAs is to reduce expression of target genes by inhibiting translation and/or reducing the concentration of mRNAs. It is believed that miRNAs regulate the expression of >60% of the protein coding genes in humans (Friedman et al., 2009).

Computational biology approaches have two broad applications in the field of miRNA studies. First, these help in identification, characterization, profiling of not only miRNAs but also target genes. Secondly, the functional significance in the context cellular regulatory systems can be analysed and studied using mathematical and stochastic simulation techniques. Recent developments in computational methods and hardware have made handling of big data easy and quicker. The advent of next generation sequencing platforms has also revolutionized miRNA studies. Now it is possible to generate genome-wide miRNA expression profiles under different conditions in a very short time with reduced cost, and key regulatory miRNAs can be easily identified. The sensitivity of this approach is far higher than traditional expression profiling techniques, including microarrays.

In this chapter, we focus on the role of computational methods in predicting miRNA gene, expression profiling the parameters that can influence the miRNA-mRNA and the influence of miRNAs in gene regulatory networks. We have also discussed our in-house computational pipelines used for miRNA expression profiling, predicting miRNA and novel miRNA from next generation sequencing (NGS) raw data.

2.1   MiRBASE

MiRBase database is an archive of miRNAs and related information from different species. It provides diverse information, including miRNAs (both precursor and mature) from different species and their target sequences. The database has three components:

   MiRBase Registry is a web interface that helps to submit information about novel miRNAs. It has created a consistent and unique nomenclature of miRNAs that prevents overlap and redundancies of gene naming. It follows an officially accepted guideline for naming a miRNA. The naming is in the form of XXXX-miR-YYY, where XXXX represents a three or four letter species code, miR or mir represents the mature and precursor hairpin respectively, and YYY represents the order of their discoveries. If two identical, mature miRNA sequences map at two different genomic loci, they are given numerical suffixes such as has-mir-6-1 and has-mir-6-2. miRNA sequence with only one or two nucleotide changes are assigned small alphabet suffixes. For example: hsa-miR-181a and hsa-miR-181b. Moreover, 5p and 3p denotes from which arm of hairpin the mature miRNA have arisen. For example, miR-142-5p and miR-142-3p are processing products from the same precursor sequence (Griffiths-Jones, 2004). A schematic representation of miRNA nomenclature is shown in Figure 2.1.

   MiRBase sequence is a primary repository of all published miRNAs and related information. It contains published mature miRNA sequences and their hairpin precursors and annotations related to their discovery, genomic locations, structure and function. Most of the annotated miRNAs are experimentally validated, and only a few are predicted sequence homologs of miRNAs verified in a related organism (Griffiths-Jones et al., 2006). Currently this database

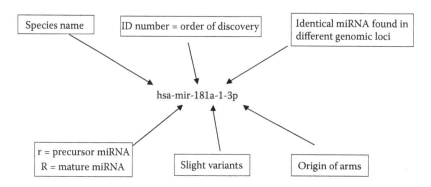

FIGURE 2.1 Schematic representation of miRNA nomenclature.

(released June 21, 2014) contains 28,645 entries from 223 species. Humans have 1,881 precursors and 2,588 mature miRNAs (http://www.mirbase.org). *MiRBase target* provides an automated pipeline for prediction of miRNA target genes. The prediction algorithm uses mature miRNA sequence to scan 3' UTR sequences from all genes of the species in Ensembl. The p-values are calculated on the basis of predicted energy of miRNA-target binding, multiple sites in a single UTR and the sites which are conserved across different species (Griffiths-Jones et al., 2006).

2.2 COMPUTATIONAL IDENTIFICATION OF miRNA GENES

Currently, 2,588 mature miRNA genes have been identified in humans (*miRBase*), and this makes up about 10% of the protein coding genes (Tyagi et al., 2008). The number is increasing every day as many new miRNA genes are being identified. Similarly, the number of reported miRNA genes of other species is also increasing at a fast rate. Computational methods have played an important role in identification of new miRNAs and rely on analysis of predicted hairpin secondary structure in the genome that resembles the stem loop of a pre-miRNA, the minimum free energy of the folded structure and the sequence conservation across the species (Akhtar et al., 2016). Validation of predicted miRNA genes is based on observation of small RNA molecules of about 22 nucleotides using northern and/or microarray, primer extension and other methods that determine the presence of expressed small RNA. In many reports the presence of small RNA in the RISC complex has been used to identify functionally relevant miRNAs. With the advent of NGS platforms, the discovery of new miRNAs has become easier. Instead of using the approach of first predicting and then validating, the new approach has been identification of all small RNAs (by small RNA sequencing) and then computational classification of only miRNAs (Akhtar et al., 2016; Gomes et al., 2013). This approach is both advantageous and limiting as it only detects expressed miRNAs. Moreover, NGS technology offers much more sensitive and reliable capabilities of identifying those lowly expressed (Wang, Gerstein, and Snyder, 2009; Mardis, 2008).

Some of the criteria used for computational miRNA gene identification in animals are as follows.

```
c     u   u        c   uc      u   c                        uagau
  acc ug ccuca gg  cagu uu ccaggaaucccu                         g
  |||  || |||||  ||  ||||  ||  ||||||||||||||                     c
  ugg ac ggagu uc  guca aa gguccuuagggg                          u
    u   u        uu      u   a                              uagaa
```

FIGURE 2.2 Representative pre-miRNA stem loop structure. (From miRBasehttp://www .mirbase.org/)

2.2.1 Sequence and Structure Conservation

Pre-miRNAs have a typical stem-loop structure as shown in Figure 2.2. Computational folding of a putative miRNA precursor sequence can be done using different approaches and then the folded structure can be compared with known structures of miRNAs. CID-miRNA uses stochastic context free grammar to fold a miRNA precursor, and hence, is a web interface to screen for a putative miRNA precursor like structure in a given sequence or genome (Wang et al., 2005; Tyagi et al., 2008). The above parameters are usually complemented by the search for conserved hairpin across the species to increase the sensitivity like miRscan and MiRseeker (Lim, Lau et al., 2003). Strong conservation suggests the existence of conserved target sites that survived the selective pressure against mutation.

2.2.2 Machine Learning Approach

Machine learning algorithms go beyond sequence and structure and use various parameters from previously verified miRNAs to predict if a given sequence is likely to be a novel miRNA. For the prediction, a set of hairpins from non-miRNA sequences is used as a negative dataset. Generally it takes into account the features, like MFE (minimum free energy of folding) of hairpin, stem region sequence conservation, loop length, occurrence of loops, mismatches and gaps on the stem region and GC content (Gomes et al., 2013). Different machine learning algorithms, such as hidden Markov model (HMM), Naïve Bayes classifier (NBC) and support vector machine have been used for miRNA gene prediction (SVM) (Mendes, Freitas, and Sagot, 2009). Other specialized machine learning methods, including CSHMM (context-sensitive Hidden Markov Model), have also been used, but with additional filters, such as presence of EST matches and Drosha cutting sites (Agarwal et al., 2010). Stochastic context free grammar (SCFG) has also been used to develop methods for prediction of miRNAs in a sequence. Some of the methods, such as the SCFG based method CIDmiRNA, can be used to scan a genomic sequence for finding miRNA genes.

2.2.3 Experimental Data-Driven Approach

Advancement in sequencing technologies has led to detection and profiling of known and novel miRNAs at an unprecedented level (Li et al., 2010: Akhtar et al., 2016). The methods generally use small RNA sequences from a next generation sequencing

TABLE 2.1

List of a Few Tools for miRNA Gene Identification

Tools	Organism	Principles	Specificity	Reference
miRscan	Human, worm	Phylogenetically conserved stem loop, resemblance to verified miRNA	70%	(Lim, Glasner et al., 2003)
RNAmicro	Animal	Conserved secondary structure, SVM based	99.47%	(Hertel and Stadler, 2006)
miRseeker	Fly	Conserved sequence, extended stem loop and pattern of nucloetide	75%	(Lai et al., 2003)
miRFinder	Human	Pre-miRNA secondary structure, SVM based	70%	(Huang, Fan et al., 2007)
PromiR II	Human, mouse, virus	Sequence, structure, Drosha cut sit, entropy, conservation, G/C ratio, free energy	96%	(Nam et al., 2006)
miRanalyser	Any	Machine learning algorithm, NGS data, miRBase	75%	(Hackenberg et al., 2009)
Microprocessor SVM	Human	Drosha processing site, structure	90%	(Helvik, Snove, and Saetrom, 2007)
CID-miRNA	Human	Secondary structure based filters, SCFG-based algorithm	–	(Tyagi et al., 2008)

platform followed by data processing and application of miRNA detection algorithms as described below. Different pipelines vary depending upon the way data is pre-processed and the nature of the miRNA detection algorithm, including the use of different folding parameters and sequence conservation (Table 2.1). The expression of miRNAs is experimentally validated by one of the methods, such as qRT-PCR, Northern hybridization and microarray technique.

2.2.4   IN-HOUSE COMPUTATIONAL PIPELINES

Here we describe our pipeline for identification and profiling of miRNAs from NGS data. During pre-processing the raw NGS data, a number of issues are handled. These include adapter and sequence trimming and removal of very short sequences. Mapping to the miRNA reference database allows us to identify all known miRNAs in terms of both number and level of expression (Figure 2.3). The reads that do not map to miRBase, or any other known coding and non-coding RNA databases, are likely to have novel miRNAs. For studying differential expression, each NGS dataset is normalized utilizing several different methods. Normalization is required to make data comparable across experiments and to reduce the impact of non-biological

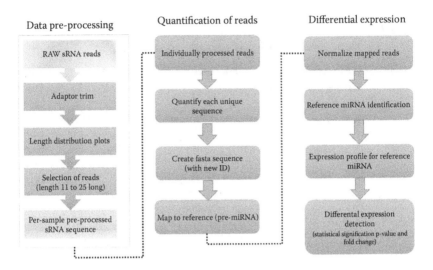

FIGURE 2.3 miRNA expression profile workflow. It is designed to estimate expression profile for reference miRNA and calculate the differentially expressed miRNA.

variability. Scaling to library size is the most commonly used normalization for RNA-Seq data sets and generally referred to as "transcript parts per million" (TPM) (Zhu et al., 2010) and "reads per kilobase per million" (RPKM). In our analysis, we have used these methods for normalization of datasets.

The expression profiles of miRNAs are useful in understanding the role of miRNA in regulation of the biological processes. There are a number of methods that have been used for expression profiling, such as northern analysis (Sempere et al., 2004), cloning (Pfeffer et al., 2005), real time polymerase chain reaction (Chen et al., 2005), microarray analysis (Thomson et al., 2004) and RNase protection assay (Lee et al., 2003). Computational methods for understanding differential expression of miRNAs also include the discovery and expression profiling of low-abundance transcripts or yet unidentified novel miRNAs. Expression profiling of reference miRNAs can be generated after alignment and normalization of the sequences. These reference miRNAs are obtained by exact matching of aligned sequences to mature miRNA database of MiRBase.

Identification of differentially regulated miRNA genes in several biological situations, including oncogenesis, has tremendous possibilities in biological and clinical research. Once datasets are normalized then these can be used for differential expression analysis using different approaches, such as the ones based on negative binomial (NB) distribution, Bayesian approach and likelihood ratio test. Using these statistical methods, various tools have been developed to identify differential expression of genes. In general, a miRNA will be significantly differentially expressed when the P-value is below a threshold (say ≤ 0.01) and at least a twofold change in normalized sequence counts.

New miRNAs can be identified from the library of small RNA sequences using a strategy shown in Figure 2.4.

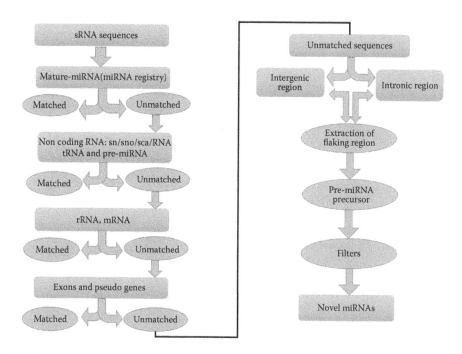

FIGURE 2.4 Flowchart for predicting a novel miRNA from small RNA sequencing data.

In principle, deep sequencing of small RNAs should generate sequences from yet unannotated regions of the genome. One can find a large number of unannotated sequences through the computational pipeline. Since miRNAs are predominantly encoded by intergenic and intronic regions, several computational programs are available to predict the novel miRNA. As the predictions are based on identifying miRNA precursors, sequences that exactly match intergenic/intronic regions, along with 70 nucleotides flanking on either side, are likely to serve as potential precursor miRNAs. We consider a precursor as potential precursor, in which the mature sequence arises from the stem portion and not from the loop part. The analysis of these sequences can be achieved with miRNA prediction algorithm tools. The accuracy of the miRNA prediction can be further improved by incorporating a probabilistic model of miRNA biogenesis to score compatibility of the position and frequency of sequenced RNA with the secondary structure of the miRNA precursor. A list of different tools that can be used for prediction of novel miRNAs is given in Table 2.2.

2.3 COMPUTATIONS IN TARGET PREDICTION

Based on computational analysis >60% of human transcript are regulated by miRNA (Friedman et al., 2009). A miRNA could be a potential target for many genes. It is a difficult task and cost prohibitive to experimentally screen the functional target of a miRNA in a complex system. Therefore, computational screening becomes the primary means to find the probable potential target gene before the experimental

TABLE 2.2

List of a Few Tools for Prediction of Novel miRNAs

#	Name	Description	Organism Specific	Prediction Features	References
1.	BioVLAB-MMIA-NGS	BioVLAB-MMIA-NGS is Cloud-based miRNA mRNA integrated analysis system using NGS data.	• Human • Mice • Plants (rice)	Structure conservation	(Chae et al., 2015)
2	CAP-miRSeq	Comprehensive analysis pipeline for deep microRNA sequencing (CAP-miRSeq) integrates read reprocessing, alignment, mature/precursor/novel miRNA qualification, variant detection in miRNA coding region, and flexible differential expression between experimental conditions.	• Any (from MiRBase)	Structure conservation	(Sun et al., 2014)
3	CPSS	CPSSis (a computational platform for the analysis of small RNA deep sequencing data), designed to completely annotate and functionally analyse microRNAs (miRNAs) from NGS data on one platform with a single data submission.	• Human • Mouse • Rat • Chicken • Fish • Other animals	Structure conservation	(Zhang, Xu et al., 2012)
4.	DARIO	A free web server for the analysis of short RNAs from high-throughput sequencing data.	• Human • Monkey • Mouse • Rat • Fly • Worm • Fish	Structure conservation	(Fasold et al., 2011)
5.	deepSOM	Novel machine learning tool for novel miRNA precursor prediction in genome-wide data.	• Animals • Plants	Sequence structure	(Stegmayer et al., 2016)
6.	eRNA	eRNA focuses on the common tools required for the mapping and quantification analysis of miRNA-seq and mRNA-seq data.	• Any	Structure conservation	(Yuan et al., 2014)
7.	HHMMiR	An approach for de-novo miRNA hairpin prediction in the absence of evolutionary conservation.	• Vertebrates • Invertebrates	Sequence structure	(Kadri, Hinman, and Benos, 2009)

(Continued)

TABLE 2.2 (CONTINUED)
List of a Few Tools for Prediction of Novel miRNAs

#	Name	Description	Organism Specific	Prediction Features	References
8.	HuntMi	HuntMi is a stand-alone machine learning miRNA classification tool.	• Animals • Plants • Viruses	Sequence structure	(Gudys et al., 2013)
9.	iMir	iMir is a modular pipeline for comprehensive analysis of small RNA-Seq data.	• Any	Structure conservation	(Gudys et al., 2013)
10.	LeARN/ smallA	A platform based on LeARN, dedicated to analyse data generated by RNA seq small RNA projects.	• Any	Structure	(Noirot et al., 2008)

work. Many pipelines have been generated to find the target gene with minimal false positive prediction. For plants, the prediction algorithms have been comparatively more effective because of their almost perfect complementarity between the plants miRNA and their targets sequence (Rhoades et al., 2002; Jones-Rhoades and Bartel, 2004). However, animal miRNAs seem to have more complex target-miRNA recognition where multiple factors are seen to be influencing the interaction. As of now, there is no perfect algorithm, hence, there is still a high rate of false positive and false negative prediction and researchers are still in search for the perfect algorithm. Here we have listed a few important parameters that influence the interaction of miRNA-mRNA complex.

1. Seed pairing/canonical sites: When it comes to target binding, the seed region (2 to 8 nts) at 5' ends of miRNAs play an important role in determining the target gene (Lewis et al., 2003; Stark et al., 2003). The six mer (seed region) sequence binds with perfect complementary to the 3' UTR of the target gene that requires little or no 3' pairing. The 7mer site contains seed region along with the eighth nucleotide of miRNA that match the target (7mer-m8 site) or seed region with A at the target position matching the 1st nucleotide of the miRNA (7mer-A1 site). The 8mer site contains the seed region flanked by both the match at eighth nucleotide of miRNA and A at the position 1 (Grimson et al., 2007; Agarwal et al., 2015). This is considered to be the most important criteria in miRNA-mRNA interaction (Grimson et al., 2007; Yue, Liu, and Huang, 2009; Lewis et al., 2003). Apparently, the GC content and the perfect pairing of the seed region has a strong correlation ($r = 0.91$), few targets pair with the canonical side when the GC content is low (Wang, 2014).

2. Non-canonical sites: The partial complementary at seed region or a bulge or wobble G:U pairing at the seed region could effectively down regulate gene expression complemented by a strong 3' paring (Brennecke et al., 2005;

Vella et al., 2004). A study has shown that disruption of the 3' pairing of miRNA, but keeping the seed region intact affects its capability to down regulate the target gene (Didiano and Hobert, 2006). The interaction with the central sites (4 to 15 nucleotides) occurs more frequently than expected and is evolutionarily conserved as shown by pull down experiments (Martin et al., 2014). Most of the miRNA-mRNA interactions show non-canonical interaction (Helwak et al., 2013; Wang, 2014; Martin et al., 2014) and generally, miRNA with low GC content in seed region follows non-canonical interaction (Wang, 2014). Tools that use this parameter are miRB Shunter and miRanda.

3. Evolutionary conservation: One of the important factors in predicting the target of miRNA with minimal false positive is to look for conservation of the target sites across the species throughout the evolution. In many cases, the orthologous sites of the complementary seed region, especially the 3' UTR, are conserved in a number of genomes (Lewis, Burge, and Bartel, 2005; Friedman et al., 2009) and the offsite seed match and 3'-compensatory site are also conserved to some extend in the target site, and this increases the sensitivity of the prediction (Friedman et al., 2009; Ghoshal et al., 2015). Around one third of the human transcripts are under selective pressure to maintain the pairing site (Friedman et al., 2009). The seed pairing site in mRNA is often flanked by adenine residues (Lewis, Burge, and Bartel, 2005). Some tools rely on non-conservation across cross-species as a filter, so this may rule out the possibility of miRNA binding sites that may not be present in closely related species. The presence of multiple, distinct and statistically significant patterns that have been seen in other miRNA binding sites in the target sequence have also been used for prediction (Miranda et al., 2006).

4. Thermodynamic energy: Many target prediction algorithms use the free energy of the duplex formed between the miRNA-mRNA complex, ΔG. If the interactions are predicted to be stable, then they are likely to be a binding partner. (Rehmsmeier et al., 2004; Ghoshal et al., 2015). To increase the sensitivity of prediction, the position and length of seed hybrid can be set, the wobble G:U base pair and the length of possible unpaired region can be specified (Kruger and Rehmsmeier, 2006).

5. The secondary structure of target site: The secondary structure of mRNA is important because it can restrict the assembly of the RISC complex and requires unfolding of mRNA for new miRNA-mRNA paring to occur (Robins, Li, and Padgett, 2005; Vella, Reinert, and Slack, 2004).

6. Site accessibility: It has been shown that insertion of ~200 bp fragment near the binding site reduces the repression equivalent to the mutation at the seed region (Kertesz et al., 2007). Moreover, the efficient binding of miRNA requires the unfolding of a local flanking target. The energy of miRNA-mRNA hybrid, that is, ΔG duplex often shows poor correlation to repression ($r = 0.36$), but taking site accessibility into consideration in this complex, ΔG, equal to the difference between the free energy gained by the binding of the miRNA to the target, ΔG duplex and the free energy

lost by unfolding the target site nucleotides, ΔG open shows high correlation to target repression (r = 0.7) (Hofacker, 2007; Ghoshal et al., 2015; Rehmsmeier et al., 2004; Kertesz et al., 2007). In mammalian systems miRNA targets are generally present in highly accessible regions, and this accessibility may be one reason why the target strength varies (Kertesz et al., 2007).

7. Proximity of the target sites: A transcript can have single to multiple binding sites for one miRNA. The number of binding sites is positively correlated to the levels of mRNA destabilization (Grimson et al., 2007). A study found that the repression of a target gene is greater when the two seed regions are close together (between 13 and 35 nts) (Grimson et al., 2007; Saetrom et al., 2007) and these sites are favorably co-conserved (Grimson et al., 2007). Therefore, it is likely that there are cooperative interactions between miRNA complexes interacting at physically close seed pairs (Saetrom et al., 2007).

8. Combinatorial effect: A combination of different miRNAs could down regulate a target gene more efficiently than that of a single miRNA, and miRNAs that are co-expressed are likely to co-regulate the target gene (Krek et al., 2005).

9. Expression database: Methods have been developed to take advantage of the existing expression databases (microarray, miR-seq, HIT-CLIP, GEO) for prediction of the target gene more accurately. For instance, incorporating the expression data of miRNA and mRNA (such as, GenMir++) an inverse correlation suggests a possible target (Huang, Babak et al., 2007). Recently, the importance of Ago in target selection is being highlighted. Tools have been developed to build and analyse 3D structures of the tertiary complex, miRNA-mRNA bound to Ago, and use a scoring system for prediction of the target (Leoni and Tramontano, 2016; Parker et al., 2009). Table 2.3 lists the tools that have been developed for prediction of the target gene based on the principles discussed above. Two or more parameters are often used in combination for accurate detection.

2.4   miRNA AND GENE REGULATORY NETWORK

It has been recognized for some time that the biological decision-making process that determines commitment of cells to different alternate possibilities is governed by the state of gene regulatory networks (GRN) (for a recent review see (Le Novere, 2015)). It is believed that a complete biological system in a given state can be defined as a vast network of all genes that interact with each other and different states or systems display different network structures. The initial studies concentrated mainly on essentially protein coding genes, and the interactions were thought to be mediated through expressed proteins that regulate the expression of these genes. However, identification of non-coding RNAs in recent years have altered our understanding of the mechanisms of gene regulation, and it has become obvious that these molecules are likely to have a profound effect on GRN. A number of ways miRNAs can, in principle, regulate GRN are outlined here.

TABLE 2.3
List of a Few Tools for miRNA Target Prediction

Tools	Organism	Principle	Specificity	Reference
PicTar	Vertibrates, fly, nematode	Seed match, conservation of seed site, free energy, combinatorial effect	76%	(Krek et al., 2005)
RNA22	Human, mouse, fly, worm	Multiple, distinct, statistically significant pattern in target site	81%	(Miranda et al., 2006)
PITA	Human, fly, mouse, worm	Site accessibility, thermodynamics	–	(Kertesz et al., 2007)
DIANA microT	Human, mouse, worm	Thermodynamic	65%	(Paraskevopoulou et al., 2013)
miRanda	Human, worm, rat, fly, mouse	Complementary binding, canonical and non-canonical binding	76%	(Enright et al. 2003, Betel et al., 2008)
miRTarBase	Any	Experimentally validated miRNA-target interaction	–	(Chou et al., 2016)
ComiR	Human, fly, worm, mouse	miRNA expression data, thermodynamic and machine learning techniques, PITA, TargetScan, miRanda	–	(Coronnello and Benos, 2013)
GenMiR++	Any	Expression dataset based on Bayesian algorithm		(Huang, Babak et al., 2007)

1. Transcription factors: If miRNAs target transcription factors (TFs) then the expression of target genes of the TF is dependent upon the level of cognate miRNAs. For example, miR-128 targets transcription factor E2F3a, an important regulator of cell cycle progression (Zhang et al., 2009). On the other hand, TFs also regulate expression of miRNA genes. For example, FosB, a TF, regulates the expression of miR-22 during PMA differentiation of K562 to megakaryocytes (Ahmad et al., 2017).

2. Epigenetics and chromatin modifications: miRNAs can target different enzymes in epigenetics and histone modification pathways, thereby modulating gene expression. Such changes are quite often related to carcinogenesis, tumor progression and response to therapy. For example, DNA methyl transferases 3A and 3B are frequently upregulated in lung cancer, and this is associated with poor prognosis. These enzymes are targeted by the miR-29 family of miRNAs, and high expression of these miRNAs are associated with better prognosis (Fabbri et al., 2007). MiR-137 post-transcriptionally represses the expression of Ezh2, a histone methyltransferase. The miR-137-mediated repression of Ezh2 feeds back to chromatin, resulting in a global decrease in histone H3 trimethyl lysine 27 (Szulwach et al., 2010). It is also possible that epigenetic changes modulate miRNA gene expression. Such changes can alter chromatin, activating oncogenic miRNAs.

For example, miR-148 is hypermethylated due to the overexpression of DNMTs in breast cancer cells, and DNA demethylation agent can reduce the tumor growth (Lujambio et al., 2008).

3. miRNAs can alter overall cellular expression machinery targeting molecules involved in protein translational, signaling pathways and post transcriptional processing.

4. Post-translational processing and stability of miRNAs: Specific gene expression is intricately linked to miRNA levels through miRNA processing pathway. Alterations in the expression of genes involved in this pathway can change the levels of mature miRNAs, thereby affecting levels of proteins whose genes are targeted by these miRNAs. For example, the Drosha binding/associated protein (like DEAD-box RNA helicases p68 (DDX5), p72 (DDX17), nuclear factor (NF) 90, and NF45), Dicer binding/associated protein (like Tar RNA binding protein (TRBP) and protein activator of PKR (PACT) and pri/pre-miRNA terminal binding protein (terminal uridylyltransferase 4) can regulate the processing of miRNA in different cell, tissue type and pathological conditions (Slezak-Prochazka et al., 2010).

It is clear from the few examples given here that miRNAs display mutual regulatory function with a number of eukaryotic gene regulatory systems. Finally, the combined network of all these systems determines the level of gene expression in a given cell. Network modules, such as feedback loop (FL) and feed forward loops (FFL) are common in GRN. Some of the examples of FFL are MYC-miR-26a-EZH2 FFL (Zhang, Zhao et al., 2012) and NF-κ B/STAT5/miR-155 (Gerloff et al., 2015). One of the systems that has been described involves miR-223 and two transcriptional factors, NFI-A and C/EBPα. The two factors compete for binding to the miR-223 promoter: NFI-A is a negative regulator of miR-223. However, when C/EBPα replaces NF1-A following retinoic acid (RA)-induced differentiation, miR-223 expression is induced. Overall it appears that an intricate regulatory network exists in controlling biological functions comprising TFs, regulatory motifs, miRNAs, epigenetic system, including histone modifications and signalling pathways. Generating and studying such complex networks is a major computational challenge.

Quantitative modelling of networks: In order to understand the role of miRNAs in modulating nature and properties of GRN and biochemical/signalling networks studies were carried out using small systems with only a few components. Two systems, a circadian oscillator model and NF-κB signalling pathway, were taken up for further studies (Nandi et al., 2009; Vaz et al., 2011). Different circadian oscillator models were used in the study and the simple one was based on an activator A and a repressor R, which are transcribed into mRNA and subsequently translated into protein. The activator A binds to the promoter and enhances the transcription rate of both A and R genes. Therefore, A is a positive element in transcription, whereas R is a negative element. miRNA was introduced in the model as a negative regulator of mRNA levels. Stochastic simulation was carried out in order to understand the behavior of the system with and without miRNA. The results showed that even in the presence of a small amount of miRNAs both frequency and amplitude of the oscillations were altered, and these changes were directly proportional to the amount of miRNAs. Both simple and complex models behaved in the same way, suggesting that observed changes in circadian oscillation in the presence of miRNAs are likely to be due to the ability of miRNAs to alter network properties.

NF-κB is involved in inflammatory response induced by pathogens. Its levels are tightly regulated since uncontrolled inflammatory response can cause serious diseases. Mathematical models have been used in understanding the dynamics and other aspects of regulation in NF-κB signalling. The effect of miRNAs on this pathway was studied using the mathematical models (Vaz et al., 2011). The motivation for this comes from studies that showed a number of miRNAs targeting different components of the NF-κB network (Taganov et al., 2006; Wang et al., 2008; Ma et al., 2011). This study considered two different situations: i) miRNAs target adaptor proteins involved in the synthesis of IKK that serves as the NF-κB activator, and ii) miRNAs target different isoforms of IκB that act as NF-κB inhibitors. The results showed that miRNAs affected the dynamics of the NF-κB signalling pathway depending on the role of the target. This "fine-tuning" property of miRNAs is essential to keep the system in check and prevents it from becoming uncontrolled.

2.5 CONCLUSION

In this chapter we have highlighted the part played by computational methods in understanding miRNA biology. These methods, in combination with next generation sequencing platforms, have helped in identification, expression and the regulatory role played by miRNAs. Further development in this area, including newer methods in complex network analysis and mathematical description of regulatory circuits, will provide a novel framework for understanding cellular decision-making processes.

REFERENCES

Agarwal S., Vaz C., Bhattacharya A. et al. 2010. Prediction of novel precursor miRNAs using a context-sensitive hidden Markov model (CSHMM). *BMC Bioinformatics.* **11 Suppl 1**: S29.

Agarwal V., Bell G. W., Nam J. W. et al. 2015. Predicting effective microRNA target sites in mammalian mRNAs. *Elife.* **4**.

Ahmad H. M., Muiwo P., Muthuswami R. et al. 2017. FosB regulates expression of miR-22 during PMA induced differentiation of K562 cells to megakaryocytes. *Biochimie.* **133**: 1–6.

Akhtar M. M., Micolucci L., Islam M. S. et al. 2016. Bioinformatic tools for microRNA dissection. *Nucleic Acids Res.* **44 (1)**: 24–44.

Betel D., Wilson M., Gabow A. et al. 2008. The microRNA.org resource: Targets and expression. *Nucleic Acids Res.* **36** (Database issue): D149–53.

Brennecke J., Stark A., Russell R. B. et al. 2005. Principles of microRNA-target recognition. *PLoS Biol.* **3 (3)**: e85.

Chae H., Rhee S., Nephew K. P. et al. 2015. BioVLAB-MMIA-NGS: microRNA-mRNA integrated analysis using high-throughput sequencing data. *Bioinformatics.* **31 (2)**: 265–7.

Chen C., Ridzon D. A., Broomer A. J. et al. 2005. Real-time quantification of microRNAs by stem-loop RT-PCR. *Nucleic Acids Res.* **33 (20)**: e179.

Chou C. H., Chang N. W., Shrestha S. et al. 2016. miRTarBase 2016: Updates to the experimentally validated miRNA-target interactions database. *Nucleic Acids Res.* **44 (D1)**: D239–47.

Coronnello C., and Benos P. V. 2013. ComiR: Combinatorial microRNA target prediction tool. *Nucleic Acids Res.* **41** (Web Server issue): W159–64.

Didiano D., and Hobert O. 2006. Perfect seed pairing is not a generally reliable predictor for miRNA-target interactions. *Nat Struct Mol Biol.* **13 (9)**: 849–51.

Enright A. J., John B., Gaul U. et al. 2003. MicroRNA targets in Drosophila. *Genome Biol.* **5 (1)**: R1.

Fabbri M., Garzon R., Cimmino A. et al. 2007. MicroRNA-29 family reverts aberrant methylation in lung cancer by targeting DNA methyltransferases 3A and 3B. *Proc Natl Acad Sci U S A.* **104 (40)**: 15805–10.

Fasold M., Langenberger D., Binder H. et al. 2011. DARIO: A ncRNA detection and analysis tool for next-generation sequencing experiments. *Nucleic Acids Res.* **39** (Web Server issue): W112–7.

Friedman R. C., Farh K. K., Burge C. B. et al. 2009. Most mammalian mRNAs are conserved targets of microRNAs. *Genome Res.* **19 (1)**: 92–105.

Gerloff D., Grundler R., Wurm A. A. et al. 2015. NF-kappaB/STAT5/miR-155 network targets PU.1 in FLT3-ITD-driven acute myeloid leukemia. *Leukemia.* **29 (3)**: 535–47.

Ghoshal A., Shankar R., Bagchi S. et al. 2015. MicroRNA target prediction using thermodynamic and sequence curves. *BMC Genomics.* **16**: 999.

Gomes C. P., Cho J. H., Hood L. et al. 2013. A review of computational tools in microRNA discovery. *Front Genet.* **4**: 81.

Griffiths-Jones S. 2004. The microRNA Registry. *Nucleic Acids Res.* **32** (Database issue): D109–11.

Griffiths-Jones S., Grocock R. J., van Dongen S. et al. 2006. miRBase: MicroRNA sequences, targets and gene nomenclature. *Nucleic Acids Res.* **34** (Database issue): D140–4.

Grimson A., Farh K. K., Johnston W. K. et al. 2007. MicroRNA targeting specificity in mammals: determinants beyond seed pairing. *Mol Cell.* **27 (1)**: 91–105.

Gudys A., Szczesniak M. W., Sikora M. et al. 2013. HuntMi: An efficient and taxon-specific approach in pre-miRNA identification. *BMC Bioinformatics.* **14**: 83.

Hackenberg M., Sturm M., Langenberger D. et al. 2009. miRanalyzer: A microRNA detection and analysis tool for next-generation sequencing experiments. *Nucleic Acids Res.* **37** (Web Server issue): W68–76.

Helvik S. A., Snove O. Jr., and Saetrom P. 2007. Reliable prediction of Drosha processing sites improves microRNA gene prediction. *Bioinformatics.* **23 (2)**: 142–9.

Helwak A., Kudla G., Dudnakova T. et al. 2013. Mapping the human miRNA interactome by CLASH reveals frequent noncanonical binding. *Cell.* **153 (3)**: 654–65.

Hertel J., and Stadler P. F. 2006. Hairpins in a Haystack: Recognizing microRNA precursors in comparative genomics data. *Bioinformatics.* **22 (14)**: e197–202.

Hofacker I. L. 2007. How microRNAs choose their targets. *Nat Genet.* **39 (10)**: 1191–2.

Huang J. C., Babak T., Corson T. W. et al. 2007. Using expression profiling data to identify human microRNA targets. *Nat Methods.* **4 (12)**: 1045–9.

Huang T. H., Fan B., Rothschild M. F. et al. 2007. MiRFinder: An improved approach and software implementation for genome-wide fast microRNA precursor scans. *BMC Bioinformatics.* **8**: 341.

Jones-Rhoades M. W., and Bartel D. P. 2004. Computational identification of plant microRNAs and their targets, including a stress-induced miRNA. *Mol Cell.* **14 (6)**: 787–99.

Kadri S., Hinman V., and Benos P. V. 2009. HHMMiR: Efficient de novo prediction of microRNAs using hierarchical hidden Markov models. *BMC Bioinformatics.* **10 Suppl 1**: S35.

Kertesz M., Iovino N., Unnerstall U. et al. 2007. The role of site accessibility in microRNA target recognition. *Nat Genet.* **39 (10)**: 1278–84.

Krek A., Grun D., Poy M. N. et al. 2005. Combinatorial microRNA target predictions. *Nat Genet.* **37 (5)**: 495–500.

Kruger J., and Rehmsmeier M. 2006. RNAhybrid: MicroRNA target prediction easy, fast and flexible. *Nucleic Acids Res.* **34** (Web Server issue): W451–4.

Lagos-Quintana M., Rauhut R., Lendeckel W. et al. 2001. Identification of novel genes coding for small expressed RNAs. *Science.* **294 (5543)**: 853–8.

Lai E. C., Tomancak P., Williams R. W. et al. 2003. Computational identification of Drosophila microRNA genes. *Genome Biol* **4 (7)**: R42.

Le Novere N. 2015. Quantitative and logic modelling of molecular and gene networks. *Nat Rev Genet.* **16 (3)**: 146–58.

Lee R. C., and Ambros V. 2001. An extensive class of small RNAs in *Caenorhabditis elegans. Science.* **294 (5543)**: 862–4.

Lee Y., Ahn C., Han J. et al. 2003. The nuclear RNase III Drosha initiates microRNA processing. *Nature.* **425 (6956)**: 415–9.

Leoni G., and Tramontano A. 2016. A structural view of microRNA-target recognition. *Nucleic Acids Res.* **44 (9)**: e82.

Lewis B. P., Burge C. B., and Bartel D. P. 2005. Conserved seed pairing, often flanked by adenosines, indicates that thousands of human genes are microRNA targets. *Cell.* **120 (1)**: 15–20.

Lewis B. P., Shih I. H., Jones-Rhoades M. W. et al. 2003. Prediction of mammalian microRNA targets. *Cell.* **115 (7)**: 787–98.

Li L., Xu J., Yang D. et al. 2010. Computational approaches for microRNA studies: A review. *Mamm Genome.* **21 (1–2)**: 1–12.

Lim L. P., Glasner M. E., Yekta S. et al. 2003. Vertebrate microRNA genes. *Science.* **299 (5612)**: 1540.

Lim L. P., Lau N. C., Weinstein E. G. et al. 2003. The microRNAs of *Caenorhabditis elegans. Genes Dev.* **17 (8)**: 991–1008.

Lujambio A., Calin G. A., Villanueva A. et al. 2008. A microRNA DNA methylation signature for human cancer metastasis. *Proc Natl Acad Sci U S A.* **105 (36)**: 13556–61.

Ma X., Becker Buscaglia L. E., Barker J. R. et al. 2011. MicroRNAs in NF-kappaB signaling. *J Mol Cell Biol.* **3 (3)**: 159–66.

Mardis E. R. 2008. Next-generation DNA sequencing methods. *Annu Rev Genom Hum Genet.* **9**: 387–402.

Martin H. C., Wani S., Steptoe A. L. et al. 2014. Imperfect centered miRNA binding sites are common and can mediate repression of target mRNAs. *Genome Biol.* **15 (3)**: R51.

Mendes N. D., Freitas A. T., and Sagot M. F. 2009. Current tools for the identification of miRNA genes and their targets. *Nucleic Acids Res.* **37 (8)**: 2419–33.

Miranda K. C., Huynh T., Tay Y. et al. 2006. A pattern-based method for the identification of MicroRNA binding sites and their corresponding heteroduplexes. *Cell.* **126 (6)**: 1203–17.

Nam J. W., Kim J., Kim S. K. et al. 2006. ProMiR II: A web server for the probabilistic prediction of clustered, nonclustered, conserved and nonconserved microRNAs. *Nucleic Acids Res.* **34** (Web Server issue): W455–8.

Nandi A., Vaz C., Bhattacharya A. et al. 2009. miRNA-regulated dynamics in circadian oscillator models. *BMC Syst Biol.* **3**: 45.

Noirot C., Gaspin C., Schiex T. et al. 2008. LeARN: A platform for detecting, clustering and annotating non-coding RNAs. *BMC Bioinformatics.* **9**: 21.

Paraskevopoulou M. D., Georgakilas G., Kostoulas N. et al. 2013. DIANA-microT web server v5.0: Service integration into miRNA functional analysis workflows. *Nucleic Acids Res.* **41** (Web Server issue): W169–73.

Parker J. S., Parizotto E. A., Wang M. et al. 2009. Enhancement of the seed-target recognition step in RNA silencing by a PIWI/MID domain protein. *Mol Cell.* **33 (2)**: 204–14.

Pfeffer S., Sewer A., Lagos-Quintana M. et al. 2005. Identification of microRNAs of the herpesvirus family. *Nat Methods.* **2 (4)**: 269–76.

Rehmsmeier M., Steffen P., Hochsmann M. et al. 2004. Fast and effective prediction of microRNA/target duplexes. *RNA.* **10 (10)**: 1507–17.

Rhoades M. W., Reinhart B. J., Lim L. P. et al. 2002. Prediction of plant microRNA targets. *Cell.* **110 (4)**: 513–20.

Robins H., Y. Li, and R. W. Padgett. 2005. Incorporating structure to predict microRNA targets. *Proc Natl Acad Sci U S A.* **102 (11)**: 4006–9.

Saetrom P., Heale B. S., Snove O. Jr. et al. 2007. Distance constraints between microRNA target sites dictate efficacy and cooperativity. *Nucleic Acids Res.* **35 (7)**: 2333–42.

Sempere L. F., Freemantle S., Pitha-Rowe I. et al. 2004. Expression profiling of mammalian microRNAs uncovers a subset of brain-expressed microRNAs with possible roles in murine and human neuronal differentiation. *Genome Biol.* **5 (3)**: R13.

Slezak-Prochazka I., Durmus S., Kroesen B. J. et al. 2010. MicroRNAs, macrocontrol: Regulation of miRNA processing. *RNA.* **16 (6)**: 1087–95.

Stark A., Brennecke J., Russell R. B. et al. 2003. Identification of Drosophila MicroRNA targets. *PLoS Biol.* **1 (3)**: E60.

Stegmayer G., Yones C., Kamenetzky L. et al. 2016. High class-imbalance in pre-miRNA prediction: A novel approach based on deepSOM. *IEEE/ACM Trans Comput Biol Bioinform.*

Sun Z., Evans J., Bhagwate A. et al. 2014. CAP-miRSeq: A comprehensive analysis pipeline for microRNA sequencing data. *BMC Genomics.* **15**: 423.

Szulwach K. E., Li X., Smrt R. D. et al. 2010. Cross talk between microRNA and epigenetic regulation in adult neurogenesis. *J Cell Biol.* **189 (1)**: 127–41.

Taganov K. D., Boldin M. P., Chang K. J. et al. 2006. NF-kappaB-dependent induction of microRNA miR-146, an inhibitor targeted to signaling proteins of innate immune responses. *Proc Natl Acad Sci U S A.* **103 (33)**: 12481–6.

Thomson J. M., Parker J., Perou C. M. et al. 2004. A custom microarray platform for analysis of microRNA gene expression. *Nat Methods.* **1 (1)**: 47–53.

Tyagi S., Vaz C., Gupta V. et al. 2008. CID-miRNA: A web server for prediction of novel miRNA precursors in human genome. *Biochem Biophys Res Commun.* **372 (4)**: 831–4.

Vaz C., Mer A. S., Bhattacharya A. et al. 2011. MicroRNAs modulate the dynamics of the NF-kappaB signaling pathway. *PLoS One.* **6 (11)**: e27774.

Vella M. C., Choi E. Y., Lin S. Y. et al. 2004. The C. elegans microRNA let-7 binds to imperfect let-7 complementary sites from the lin-41 3'UTR. *Genes Dev.* **18 (2)**: 132–7.

Vella M. C., Reinert K., and Slack F. J. 2004. Architecture of a validated microRNA: Target interaction. *Chem Biol.* **11 (12)**: 1619–23.

Wang H., Garzon R., Sun H. et al. 2008. NF-kappaB-YY1-miR-29 regulatory circuitry in skeletal myogenesis and rhabdomyosarcoma. *Cancer Cell.* **14 (5)**: 369–81.

Wang X. 2014. Composition of seed sequence is a major determinant of microRNA targeting patterns. *Bioinformatics.* **30 (10)**: 1377–83.

Wang X., Zhang J., Li F. et al. 2005. MicroRNA identification based on sequence and structure alignment. *Bioinformatics.* **21 (18)**: 3610–4.

Wang Z., Gerstein M., and Snyder M. 2009. RNA-Seq: A revolutionary tool for transcriptomics. *Nat Rev Genet.* **10 (1)**: 57–63.

Yuan T., Huang X., Dittmar R. L. et al. 2014. eRNA: A graphic user interface-based tool optimized for large data analysis from high-throughput RNA sequencing. *BMC Genom.* **15**: 176.

Yue D., Liu H., and Huang Y. 2009. Survey of computational algorithms for microRNA target prediction. *Curr Genomics.* **10 (7)**: 478–92.

Zhang X., Zhao X., Fiskus W. et al. 2012. Coordinated silencing of MYC-mediated miR-29 by HDAC3 and EZH2 as a therapeutic target of histone modification in aggressive B-Cell lymphomas. *Cancer Cell.* **22 (4)**: 506–23.

Zhang Y., Chao T., Li R. et al. 2009. MicroRNA-128 inhibits glioma cells proliferation by targeting transcription factor E2F3a. *J Mol Med (Berl)*. **87 (1)**: 43–51.

Zhang Y., Xu B., Yang Y. et al. 2012. CPSS: A computational platform for the analysis of small RNA deep sequencing data. *Bioinformatics*. **28 (14)**: 1925–7.

Zhu E., Zhao F., Xu G. et al. 2010. mirTools: MicroRNA profiling and discovery based on high-throughput sequencing. *Nucleic Acids Res*. **38** (Web Server issue): W392–7.

3 miRNAs: Small RNAs with Big Regulatory Functions in Parasitic Diseases

Sneha Anand and Rentala Madhubala

CONTENTS

Introduction

MicroRNAs (miRNAs) are a subclass of small regulatory RNAs, expressed by almost all metazoans. They regulate genetic expression either by direct cleavage or by translational repression of the target mRNAs recognized through partial complementary base pairing. The active and functional unit of miRNA is its complex with Argonaute proteins known as the microRNA-induced silencing complex (miRISC). They have been known to regulate various developmental and physiological processes. Dysregulated miRNA expression in the human cell has been linked to a diverse group of disorders like cancer, cardiovascular dysfunctions, liver damage, immunological dysfunction, metabolic syndromes and pathogenic infections. An increasing number of studies has revealed that miRNAs are indeed a pivotal component of host-pathogen interactions and play an important role in host immune responses toward microorganisms. miRNA is emerging as an important tool for genetic study, therapeutic development and diagnosis for human pathogenic infections caused by various pathogenic organisms like viruses, bacteria, parasites and fungi. Many pathogens have been known to exploit the host miRNA system for their own benefit, including pathogenesis, survival inside the host cell and by-passing some host immune barriers. Other pathogens express their own miRNA inside the host, contributing to their replication, survival and/or latency. This chapter aims to highlight the role and significance of miRNA in relation to some important human parasitic diseases.

3.1   miRNAs: SMALL RNAs WITH BIG REGULATORY FUNCTIONS

miRNAs are endogenous small noncoding RNAs that regulate gene expression by binding to target messenger RNAs and inducing their translational repression, cleavage, or accelerated decay (Bartel 2004). miRNAs are produced by an elaborate processing mechanism. Their tissue-specific or developmental stage-specific expression indicates that their cellular expression is tightly regulated. Protozoan miRNAs do not share significant homology to known miRNAs of plants and animals. The genes encoding the RISC machinery (Dicer, Argonaute and Piwi) in protozoan parasites are grouped in a distinct phylogenetic lineage from metazoans (Bartel 2004; Zheng et al. 2013; Bayer-Santos et al. 2017). In addition, protozoans also lack Drosha (a nuclear RNaseIII) homologs, suggesting the presence of a Drosha-independent pathway in these parasites. RNA-dependent RNA polymerase like-genes, which encode an enzyme implicated in the amplification of pre-miRNAs, have been identified in *Giardia lamblia*, *Entamoeba histolytica*, and *Toxoplasma gondii* (Bayer-Santos, Marini et al. 2017). RISC components like Argonaute, Piwi, Dicer, RNAse III and miRNAs have been identified in *Trypanosoma brucei*, *Trypanosoma congolense*, *Leishmania braziliensis*, *T. gondii*, *Neospora caninum*, *G. lamblia*, *Trichomonas vaginalis* and *E. histolytica*, suggesting that these organisms have a classical RNAi pathway (Militello et al. 2008; Atayde et al. 2011; Hakimi and Cannella 2011; Zheng et al. 2013). Extensive bioinformatics and functional analyses have shown that *Trypanosoma rangeli*, *Leishmania donovani*, *Leishmania major*, *Trypanosoma cruzi*, and *Plasmodium falciparum* are RNAi-deficient organisms (Robinson and Beverley 2003; DaRocha et al. 2004; Baum et al. 2009; Atayde et al. 2011; Stoco et al. 2014; Bayer-Santos et al. 2017). It is observed that orthologs of RNAi machinery are present only as pseudogenes in *T. rangeli*, *L. major*, *Leishmania infantum*, and *Leishmania mexicana* (Zheng et al. 2013; Bayer-Santos et al. 2017). It is important to carry out further functional analysis in order to confirm these *in silico* observations, because a mere failure to identify protein homologs using bioinformatics in genome databases does not imply that the protein function was in fact lost. In these cases, there is an equal possibility that the primary sequences diverged too much to be detected by sequence homology. Further studies are required to identify a full RNA repertoire of these parasites.

In vertebrates it is known that miRNA can regulate the expression of several different genes. They are known to function as "master-switches" that can efficiently coordinate multiple cellular pathways and processes. Underlining their importance in mammals, miRNA misregulation has been associated with the onset, development and pathophysiology of many diseases including cancer. It is now known that when cells encounter microorganisms, miRNA-based regulatory pathways are also perturbed that contribute to the mounting of host cell responses/defenses. The importance of miRNAs has been described in several parasitic pathologies. The following sections summarize our current understanding of miRNAs in parasitic diseases caused by both intracellular and extracellular parasites.

3.2 INTRACELLULAR PROTOZOAN PATHOGENS

Invasion and replication inside a mammalian cell is a common strategy used by protozoan pathogens to evade the host immune system. In order to survive in the intracellular environment protozoan parasites modulate the expression of host miRNAs associated with different biological processes. Conversely, host miRNAs are also known to inhibit the proliferation of microorganisms by targeting virulence and essential genes of the parasite. Thus, any change in miRNA profiles might indicate either a subversion strategy by the parasite or a defense mechanism by the cell (Hakimi and Cannella 2011). The role of miRNAs in infections caused by some intracellular parasites is outlined below.

3.2.1 ALTERATION OF HOST miRNAs BY APICOMPLEXANS

Plasmodium spp., *T. gondii*, and *Cryptosporidium parvum* are medically important intracellular parasites that belong to the phylum Apicomplexa. As obligate intracellular parasites, they use different mechanisms to reorganize host cell functions to allow their survival and replication inside host cells (Plattner and Soldati-Favre 2008). Considerable evidence suggests the role of miRNAs in infections caused by apicomplexans. Although miRNA-silencing mechanisms are absent in some apicomplexans, a number of recent studies have reported the role of miRNAs in multiple host-parasite interactions. Once inside, parasites regulate host gene expression to improve their abilities to infect and proliferate inside host cells. This usually occurs via inhibition of host immune responses, especially those involved in apoptosis and cytokine production (Plattner and Soldati-Favre 2008). A growing body of evidence has demonstrated that parasites modify host miRNA profile, underscoring the importance of miRNAs in host-parasite interactions. The alterations induced on host miRNAs by apicomplexan parasites like *Plasmodium*, *Cryptosporidium*, and *Toxoplasma* has been summarized below.

One of the early examples known is that of the role of miRNA in *Cryptosporidium* infections. It is known that both innate and adaptive immune responses are essential for resolution of cryptosporidiosis and also resistance to subsequent infections (Akira and Takeda 2004; Chen et al. 2005). *C. parvum* infection modulates several miRNAs involved in the Toll-like Receptor 4 (TLR-4) and NF-κB signaling pathways (Figure 3.1a).

Using an *in vitro* model of human cryptosporidiosis, Zhou et al. (2009) showed that *Cryptosporidium* infection of human cholangiocytes decreased expression of the let-7 miRNA via a NF-κB-dependent mechanism. This downregulation was further linked with the upregulation of TLR-4 (let-7 target), resulting in a better epithelial defense response against the parasite (Chen et al. 2007). Additionally, let-7 together with miR-98 also alters NF-κB signaling by regulating the expression of a negative regulator of inflammatory cytokine signaling (SOCS/CIS) (Hu et al. 2010). Induction of SOCS/CIS enhances IkBα degradation resulting in NF-κB transcription factor activation. Conversely, the infection was also able to induce expression of a series of miRNAs like miR-125b, miR-21, and miR-30b. The induction of these miRNAs has been associated with the better survival of the parasite inside the host cells.

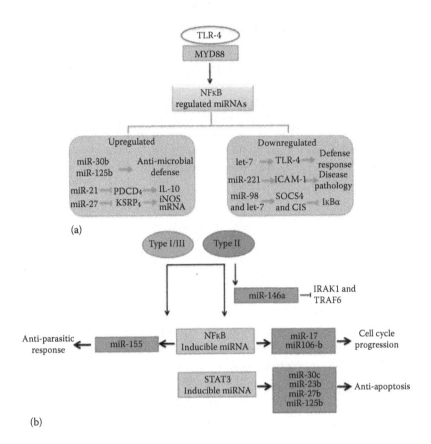

(a)

(b)

FIGURE 3.1 (a) *Cryptosporidium* infection alters NF-κB-regulated miRNAs. Upregulation of miR-30b and miR-125b results in induction of an antimicrobial defense response. Induction of miR-21 blocks the pro-inflammatory protein PDCD4, leading to suppression of interleukin 10 (IL-10). MiR-27 induction blocks expression of KSPR₄ further modulating iNOS levels. On the other hand, NF-κB mediated downregulation of let-7 causes upregulation of TLR-4 (let-7 target) resulting in an anti-parasitic defense response. Decreased miR-221 expression controls ICAM-1 thus controlling disease pathology. Downregulation of let-7 along with miR-98 control the expression of signaling proteins SOCS/CIS. Induction of SOCS/CIS enhances IκBα degradation resulting in activation of NF-κB regulated inflammatory genes. TLR4 (Toll-Like Receptor); MyD88 (Myeloid Differentiation primary response gene 88); PCDP4 (Programmed Cell Death Protein 4); KSRP4 (KH-type splicing regulatory protein 4); ICAM-1 (Intercellular Adhesion Molecule 1); iNOS2 (inducible Nitric Oxide Synthase 2). (b) *Toxoplasma* infection controls miRNA expression via NF-κB signaling and STAT3 transactivation. Type I, II, and III strains induce expression of miR-155 whereas the type II strain shows an exclusive and significant induction of miR-146a, a key immune and inflammatory response regulator targeting IRAK1 and TRAF6. Both miR-155 and miR-146a are regulated by NF-kB signaling. Other miRNAs regulated by NF-κB signaling include miR-17 and miR-106b that are known to play crucial roles in cell cycle regulation. STAT3 binding was demonstrated to regulate a subset of miRNAs (miR-30c, miR-125b-2, miR-23b, miR-27b,) that were induced by *T. gondii* infection in human macrophages. These miRNAs were shown to influence anti-apoptotic pathway.

For example, induction of miR-21 expression after *C. parvum* infection results in negative feedback regulation of TLR-4/NF-κB signaling, as miR-21 targets PDCD4, a pro-inflammatory protein that promotes activation of NF-κB and suppresses interleukin 10 (Sheedy, Palsson-McDermott et al. 2010).

Other NF-κB responsive miRNAs that are altered during *Cryptosporidium* infection include miR-27b and miR-221. MiR-27b expression is upregulated during infections, thus blocking the expression of its target mRNA KSRP4. Downregulation of KSRP4 modulates iNOS levels (Zhou, Gong et al. 2012) contributing to parasite survival. *C. parvum* infection also results in decreased miR-221 expression in infected epithelial cells. Since intercellular adhesion molecule-1 (ICAM-1) is described as the direct target of miR-221, downregulation of miR-221 is probably involved in increased infiltration of lymphocytes into the intestinal mucosa (Gong et al. 2011).

Global miRNA expression profiles for host cells infected with *Toxoplasma* have also shown interesting results. It has been shown that the parasite specifically modulates expression of important host miRNAs during infection (Figure 3.1b) (Zeiner et al. 2010). After 24 h, *T. gondii* infection altered around 14% of host miRNAs in primary human foreskin fibroblasts that further resulted in activation of NF-kB signaling (Shapira et al. 2002). Primary transcripts for miR-17~92 and miR-106b~25 that are known to play crucial roles in mammalian cell cycle regulation were also found to be upregulated. Both of these miRNAs have been shown to influence the functionally intertwined pathways of apoptosis and G1/S cell cycle progression (Xiao and Rajewsky 2009). It has also been demonstrated that *Toxoplasma* infection in human macrophages regulates a subset of miRNAs (miR-125, miR-30c, miR-125, and miR-19a/b) that is controlled by STAT-3 signaling. These miRNAs are mainly involved in anti-apoptosis in response to *T. gondii* infection (Cai et al. 2013). Recent studies have also linked two immunomodulatory miRNAs-miR-155 and miR-146a-to *T. gondii* infection. MiR-146a is a key immune response regulator targeting TRAF6 and IRAK1 (Saba et al. 2014). Both these miRNAs are involved in control of parasitic burden in the host.

A correlation between plasma miRNA levels and *T. gondii* infection has also been observed (Jia et al. 2014). When miRNA expression profiles from *T. gondii*-infected mice were compared with corresponding healthy mice. Among all the modulated miRNAs, three miRNAs (miR-712, miR-511, and miR-217) were found to be upregulated. Moreover, the upregulation of these miRNAs was shown to be a specific response to *T. gondii* infection, as challenge with other pathogens such as *Plasmodium berghei*, *P. yoelii*, *P. chabaudi*, and *C. parvum*, resulted in downregulation of these miRNAs. The parasite-specificity of miR-712, miR-511, and miR-217 make them good biomarkers for *T. gondii* infection.

Another very important intracellular parasite is *Plasmodium*. Significant material exchange occurs between the host cell and *Plasmodium* parasite during the intra-erythrocytic developmental cycle. Interestingly Xue et al. detected the presence of human miRNA within the parasite; however, they were unable to find any parasite-specific miRNA (Xue et al. 2008). This is in agreement with previous reports, which have failed to identify the main components of RISC and Dicer complex in *P. falciparum* genome (Coulson et al. 2004; Hall et al. 2005). Thus, due to the absence of its own machinery probably the parasite imports human miRNAs for its own functions. Of the 132 short

RNA found by Xue et al. (2008), 54 were rRNAs and tRNAs from human blood and *Plasmodium*, 18 were degraded fragments of human blood and *Plasmodium* mRNA, 26 were human miRNA, and 24 did not match the human or *Plasmodium* genome.

Different studies have observed high expression of miR-451 in parasitized RBC (pRBC) as compared to uninfected normal RBC. However, a better understanding of the role of miR-451 in *Plasmodium* infections came from a study using sickle cell (HbS) erythrocytes. In this study the authors were able to identify the role for miRNAs from HbS erythrocytes in resistance against malaria (LaMonte et al. 2012). This is the first report in which the translocation of human miRNAs into the parasite was characterized. Using multiplex real time PCR, presence of ~100 human miRNAs were detected within the parasites. Their findings reveal that the translocation of human miRNAs into *P. falciparum* is a defense mechanism in erythrocytes and an important factor of malaria resistance in sickle cell erythrocytes. Although erythrocytes do not contain nuclei, these cells have abundant levels of miRNAs. Notably, the expression of miR-451 is highly upregulated during erythropoiesis, resulting in high levels in mature erythrocytes (Masaki et al. 2007; Nelson et al. 2007).

When the miRNA uptake profile across intra-erythrocytic developmental cycle was monitored, it was shown that the expression of miR-451 peaked after 32 h, indicating a dynamic uptake of miRNAs during this time period. The authors in this study also provide evidence that miR-451 forms chimeric fusions with *P. falciparum* mRNAs. It was observed that miR-451 integrates into *P. falciparum PKA-R* transcripts leading to reduced translation of the regulatory PKA subunit, and thus, resulting in increased PKA catalytic activity. Precise regulation of PKA activity is essential during several stages of parasite development, including erythrocyte invasion and survival, sporozoite motility, hepatocyte invasion as well as induction of gametocytogenesis (Trager, Gill et al. 1999; Li and Cox 2000; Ono, Cabrita-Santos et al. 2008). Hence, disruption of this pathway by the host represents an excellent opportunity to disrupt parasite development. Correlating this with sickle cell diseases, they observed that two miRNAs (let-7i and miR-451) that negatively modulate parasite growth were enriched in sickle cell erythrocytes. Thus, enhanced expression of miRNAs contributes to malarial resistance of both HbAS and HbSS erythrocytes. This study provided the first data on human miRNAs regulating *Plasmodium* gene expression and suggested the possibility of miRNAs being incorporated into malaria parasites. Moreover, a very recent study investigated plasma miRNAs alterations mediated by *P. vivax* and showed downregulation of miR-451 and miR-16 in *P. vivax* malaria infection (Chamnanchanunt et al. 2015).

It was recently shown that erythrocyte miRNAs also play an important role in communication between host cells during *Plasmodium* infection. Infected erythrocyte releases extracellular vesicles (EVs) carrying functional miRNAs (Argonaute 2 complexes, enriched with miR-451 and let-7b) in the bloodstream. Endothelial cells internalize these EVs and the delivered miRNAs change recipient cell gene expression, culminating in altered endothelial cell barrier properties (Mantel, Hjelmqvist et al. 2016).

The expression profile of miRNAs has also been evaluated in experimental malaria models. When modifications in liver miRNAs were investigated in mice infected with self-healing *P. chabaudi* malaria (Delic, Dkhil et al. 2011), it was observed that primary infections could induce an immune response, which promoted the alterations in liver miRNAs. Changes in hepatic miRNAs usually associated with adaptive immune responses were detected: miR-26b, MCMV-miR-M23-1-5p, and miR-1274a were found upregulated and 16 miRNA species (miR-101b, let-7a, let-7g, miR-193a-3p, miR-192, miR-142-5p, miR-465d, miR-677, miR- 98, miR-694, miR-374, miR-450b-5p, miR-464, miR-377, miR- 20a, and miR-466d-3p) were down-regulated. However, the expression level of these miRNAs (related to the immune response) remained unchanged for almost all of them in re-infected mice. Thus, the changes in the miRNA levels were found to correlate with the healing of the parasitic infection and therefore, production of a potent adaptive immune response against the secondary infections. Although the data did not explain the mechanisms underlying the changes in miRNAs expression, they appear specific for malaria infection and important in acquired protective immunity against *P. chabaudi*. A recent report has also suggested significant upregulation of miR155 in the liver after infection with genetically attenuated parasites (GAP) (Hentzschel, Hammerschmidt-Kamper et al. 2014). Additionally, GAP injection also induced TNF-α and IFN-γ expression, two known upstream regulators of miR-155, thus underscoring the importance of miRNA in initiation of a potent immune response against the parasite.

The pathogenesis of cerebral malaria is multifaceted and evidence suggests that the host immune system plays a major role in the pathophysiology of this disease. It has been seen in previous studies that immune modulation (Hunt and Grau 2003), apoptosis (Lackner et al. 2007), leukocyte cytoadhesion (Costa et al. 2011), and hypoxia (Penet et al. 2005) are also involved in pathogenesis of this disease. El-Assad et al. compared expression levels of selected miRNAs involved in immune modulation, apoptosis, leukocyte cytoadhesion and hypoxia (El-Assaad, Hempel et al. 2011). In this experiment, the miRNA profile of mice infected with *P. berghei* ANKA (cerebral malaria strain) was compared to mice infected with non-cerebral malaria strains (Pb- K173 or PbK). Specific miRNAs like let-7i, miR-27a, and miR-150 were found to be upregulated in brain tissue of *P. berghei* ANKA infected mice. While let-7i has been described to control cellular proliferation and the innate immune response (O'Hara et al. 2010), miR-150 is highly expressed in monocytes and is related to cell proliferation and apoptosis. This study also showed that miR-150 is involved in controlling monocyte accumulation in microvasculature, which is one of the key features of fatal cerebral malaria. Increased expression of miR-27a was found to be associated with apoptosis, increased TNF sensitivity, regulation of T cell proliferation and the NF-κB signaling pathway during inflammation (Chhabra et al. 2009; Tourneur and Chiocchi 2010). Modulation of these miRNAs in cerebral malaria suggests specific roles for these miRNAs in the neurological cerebral malaria syndrome. Recently it was shown that miR-155 also has an important role in the pathogenesis of cerebral malaria. MiR-155 was found to negatively regulate blood-brain-barrier integrity and T cell function. In a mouse experimental model of cerebral malaria, the authors demonstrated that blocking miR-155 function by gene knockout or pre-treatment with miR-155 antagomir enhances endothelial quiescence, bloodbrain-barrier integrity and host survival (Barker et al. 2017). Thus further investigations would be useful in exploring the potential therapeutic use of miR-155 in the treatment of cerebral malaria.

Taken together, these studies highlight the importance of miRNAs in *Plasmodium* infection and strongly suggest that a reprogramming of miRNA expression could have a regulatory function in malaria pathogenesis.

3.2.1.1 Modulation of Host microRNome by Kinetoplastids

The kinetoplastids are flagellated unicellular eukaryotic organisms. This group includes parasites responsible for a number of diseases in humans and animals. The most relevant organisms causing disease in humans are *Trypanosoma cruzi* and *Leishmania* sp. One of the main characteristics of the members of the kinetoplastids is their ability to live as obligate intracellularparasites within different cell types, including macrophages, which are the main immune cells acting as the first line of host defense.

It is known that *T. cruzi* releases different populations of extracellular vesicles (EVs) that contain different sets of small RNAs (Bayer-Santos et al. 2013, 2014; Fernandez-Calero et al. 2015). These vesicles are transferred between *T. cruzi* cells, increasing their differentiation rate, and also between parasites and mammalian host cells, inducing changes in gene expression (Garcia-Silva, das Neves et al. 2014). In recent studies it was also shown that mammalian cells incubated with *T. cruzi* can release EVs which bind to trypomastigotes and protect them against lysis by the complement system (Cestari et al. 2012; Ramirez et al. 2017).

Inflammatory lesions have been observed in patients with chronic Chagas disease cardiomyopathy (CCC) and in murine models having *T. cruzi* infections. Previous studies have observed that *T. cruzi* EVs carry virulence factors and that treatment of mice with them can induce severe heart pathology and inflammation even before infection. In these studies high levels of pro-inflammatory cytokines and nitric oxide were detected, suggesting that EVs are involved in disease related pathology (Torrecilhas et al. 2012; Nogueira et al. 2015). Altered miRNA profile of patients with CCC as compared to idiopathic dilated cardiomyopathyalso suggested a role of parasitic miRNA in regulation of CCC pathogenesis (Ferreira, Frade et al. 2014; Navarro, Ferreira et al. 2015; Linhares-Lacerda and Morrot 2016). Thus, it could be possible that miRNAs are released from the parasite as extracellular vesicles. Currently there are no reports available suggesting the presence of any such mechanism in Trypanosomes. However, a recent report has suggested that the exosomes in Trypanosomes are formed in multivesicular bodies (MVB) utilizing a mechanism similar to microRNA secretion in mammalian cells (Eliaz, Kannan et al. 2017). Studies in murine model have shown that *T. cruzi* infection can also induce changes in miRNAs levels of thymic epithelial cells (Linhares-Lacerda and Morrot 2016). *Leishmania donovani* and *L. braziliensis* are also reported to release EVs containing RNAs, suggesting that the packaging of specific RNA sequences into EVs may be a conserved phenomenon (Lambertz et al. 2015).

Additional work has also shown that infection of macrophages (Lemaire, Mkannez et al. 2013; Frank, Marcu et al. 2015) and dendritic cells (Geraci, Tan et al. 2015) by *Leishmania* changes host cell miRNA expression profile. However, the mechanisms regulating such changes have not been clearly determined in these studies. *Leishmania donovani*, can take advantage of miRNAs to regulate lipid metabolism. High serum cholesterol causes resistance to *L. donovani* infection, while *L. donovani* infected liver contains low levels of cholesterol and altered expression of lipid metabolic genes, many of which are direct or indirect targets of miR-122.

Glycoprotein GP63, a Zn-metalloprotease on the surface of *L. donovani*, targets pre-miRNA processor DICER1 to prevent miRNA production, including miR-122. Hence, *L. donovani* exploits the host liver metabolism to aid in infection by suppressing Dicer and down-streaming miRNAs that lower cholesterol levels (Ghosh et al. 2013). A recent report has also identified the regulatory role of the miR30 family member, miR30A-3p as the primary modulator of host cell autophagy after infection with *L. donovani*. It was demonstrated in this study that BECN1/Beclin 1, a key autophagy-promoting protein, is a potential target of miR30A-3p, which negatively regulates BECN1 expression. Enhanced expression of MIR30A-3p during *L. donovani* infection, suggests a regulatory role of this miRNA in the modulation of host cell autophagy (Singh et al. 2016).

A recent report by Bose et al. put forward a new mechanism to explain the dysregulation of host cell miRNA expression by *L. donovani*. They proposed that *L. donovani* infection blocks the maturation of endosomes to late endosome by targeting the endosomal protein HRS (hepatocyte growth factor regulated tyrosine kinase substrate). Downregulation of HRS in *L. donovani*–infected macrophages prevents uncoupling of mRNA-AGO interaction, blocking degradation of translationally repressed messages and recycling of miRNPs. Correlating this with the immunology of the disease they show that due to this uncoupling, let-7a miRNPs remain bound to target mRNAs and fail to repress newly formed IL-6 mRNA, thereby, enhancing the translation of IL-6 in host cells. Production of IL-6 then further helps the parasite to suppress host macrophage activation and promote infection (Bose et al. 2017).

3.3   EXTRACELLULAR PROTOZOAN PARASITES

Interaction between extracellular protozoan parasites and host cells is critical for tissue adhesion, colonization and damage. *E. histolytica* and *T. vaginalis* are two medically important extracellular parasites that live in the gastrointestinal and urogenital tracts, respectively. To establish a focus of infection it is important for these parasites to adhere to epithelial cells. *T. vaginalis* trophozoites express adherence factors that allow parasite attachment to ectocervical epithelial cells. Research by Twu et al. (2013) showed that trophozoites release exosomes carrying small RNAs and parasite-specific proteins that fuse with host cells to deliver their cargo. It was shown by them that *T. vaginalis* exosomes modulate the secretion of pro-inflammatory cytokines like IL-6 and IL-8, thus allowing the parasite to sustain a chronic infection (Twu et al. 2013). It was proposed that *T. vaginalis* infection controls the expression of these cytokines by modulating host miRNAs. The study also suggested that EVs released from a highly adherent strain of *T. vaginalis* were able to enhance attachment of a poorly adherent strain to both vaginal and prostate epithelial cells.

Modulation of the host cells miRNA profile also occurs in amebic infections. *E. histolytica* kills intestinal epithelial cells after attachment of trophozoite to host cell membranes by inducing Ca2+ signaling that culminates in cell death. This is followed by a potent inflammatory response and invasion of colonic crypts by trophozoites. Marie et al. used a genome-wide miRNA library to screen human epithelial cells susceptible to killing by *E. histolytica*. Thus, using this, they were able to identify specific host factors required for amebic cytotoxicity (Marie, Verkerke et al. 2015).

Trypanosoma brucei is an extracellular parasite that is transmitted to a mammalian host by an insect vector (tsetse fly). In both hosts, they are extracellular and migration to specific host tissues is essential for parasite development and pathogenesis. Bloodstream trypomastigotes can move out of blood vessels and invade extravascular tissues including the central nervous system. In tsetse fly, epimastigotes attach to epithelial cells in the salivary gland and differentiate into metacyclic trypomastigotes, which are infective for humans. Recently, *T. brucei* was reported to produce EVs (Szempruch, Sykes et al. 2016). EVs produced by *T. brucei rhodesiense* contain the serum resistance-associated protein (SRA) that is necessary for human infectivity. In addition, *T. brucei* EVs have also been shown to fuse with mammalian erythrocytes, altering the physical properties of the membrane and causing erythrocyte lysis and anemia (Szempruch et al. 2016b). Despite this interesting work, there is no report of transference of non-coding RNAs in *T. brucei* to date (Szempruch, Sykes et al. 2016). However, a recent report has suggested that the exosomes in *T. brucei* are formed in MVB utilizing a mechanism similar to microRNA secretion in mammalian cells (Eliaz, Kannan et al. 2017).

3.4 MICRORNAs IN PARASITIC HELMINTHIASES

Parasitic helminths, belonging to the phyla Nemathelminthes (nematodes) and the Platyhelminthes (including trematodes and cestodes) are the most common parasites affecting humans.

Lack of coding genes for the key miRNA processing enzymes like Dicer and Argonaute has been a distinctive feature of several protozoan parasites (e.g., *Plasmodium* spp.), thus indicating exploitation of host miRNA pathways by these parasites. In contrast to these unicellular organisms, it has been shown that the key proteins involved in miRNA processing are conserved across different parasitic helminths (Gomes et al. 2009; Dalzell et al. 2011; Zheng et al. 2013; Hoogstrate et al. 2014; Cai et al. 2016). There is considerable data available regarding miRNA profiles of parasitic helminths and alterations of host miRNAs in response to helminth infection. Using miRNA prediction tools, different miRNA populations have been identified in around 35 species of parasitic helminths.

The complex life cycles of parasites reflect their constant need to respond to changing environmental conditions. For example, there are at least seven discrete developmental stages associated with the schistosome life cycle. Thus, their life cycle involves dramatic morphological and transcriptional changes within their intermediate and definitive hosts (Fitzpatrick, Peak et al. 2009; Gobert, Moertel et al. 2009). It is known that in the free-living parasite *C. elegans*, let-7 miRNA plays a key role in the heterochronic pathway that controls the temporal patterning of larval development (Rausch et al. 2015). The let-7 family in *C. elegans* has also been shown to modulate innate immune responses during pathogen stress, thus, ensuring robust timing of developmental events (Sun et al. 2016). Therefore, based on the observation that different miRNAs are expressed at various developmental stages in helminth parasites (Cai et al. 2011, 2013), specific regulatory roles for these parasitic miRNAs have been elucidated.

miRNAs play key regulatory roles in host–pathogen interaction. It is known that after parasitic infection, host miRNAs are dysregulated, hinting there subtle but crucial role in regulation of pathogen-associated pathology. Thus, during infection with different parasitic helminthes high-throughput methods like deep sequencing and miRNA microarray have been used to explore host miRNA profiles in pathology-relevant tissues. Some examples of these have been described below.

In case of *S. japonicum* infection, differential expression of host miRNAs was observed in different tissues (e.g., lung, spleen, and liver) in a murine model (Han, Peng et al. 2013). Consistent with the migratory route of developing schistosomula, the lung is one of the most susceptible tissues at the early stage of infection. This is reflected in the modulation of a relatively high number of miRNAs in lung tissue. Some of these miRNAs (e.g., miR-125a-5p, miR-200a, miR-181a/b/c, and miR-150) are known to be key regulators of various immune responsive genes, thus implying that innate immune responses and the wound-healing process are important functions requiring modulation in the host lungs during the early stages of *S. japonicum* infection (Burke, McGarvey et al. 2011). During an active stage of *schistosome* infection temporal analysis of the host miRNA in liver suggested that a unique panel of miRNAs was dysregulated (Cai et al. 2013; Hoy et al. 2014; He et al. 2015). Again, in this case, it was seen that the immune cell-specific miRNAs (e.g., miR-155, miR-146a/b, and miR-223) were predominantly upregulated (Cai et al. 2013). This implied that the presence of miRNAs act as molecular signatures for the infiltration of immune cells (e.g., neutrophils) and activation of residential cells in the liver that further results in the development of schistosomal hepatopathology (Burke et al. 2010; He et al. 2013). Thus, these studies highlight the regulatory role of miRNAs in tuning the immune response during parasitic infection.

Further reports have also indicated that the development of certain tissue related pathologies might also alter miRNA levels in the wider circulatory system. It has been hypothesized that miRNAs may be passively released during cell necrosis or actively secreted in vesicular structures such as exosomes. These cell-free miRNAs can be stably detected in a wide range of body fluids, like urine, blood plasma or serum (Chen et al. 2008; Mitchell et al. 2008). Three factors are responsible for the high stability of miRNAs in biofluids: (i) their incorporation into exosomes or other extracellular vesicles (EVs); (ii) the formation of a miRNA–protein complex; (iii) and their small size (Zhu et al. 2014). Circulating miRNAs have been regarded as promising diagnostic biomarkers in various diseases including liver damage, cancers and viral infections (Kosaka, Iguchi et al. 2010; Starkey Lewis, Merz et al. 2012). In case of *S. japonicum* infection, host circulating miRNA, miR-223 has been suggested as a promising biomarker (Kosaka, Iguchi et al. 2010). However, the serum level of miR-223 has also been reported to be upregulated in other types of disease, such as sepsis (Wang, Yu et al. 2010), indicating the limitation of using limited miRNA signature as a diagnostic biomarker. Thus, indicating that a panel of miRNAs may, therefore, better guarantee the specificity of diagnosis for parasitic diseases. Currently, miRNAs within biological fluids are being studied as good disease-related markers and have emerged as a powerful tool for solving many difficulties in both diagnosis and treatment. During pathological conditions, alteration of a specific miRNA, compared to a healthy control, can be used as a biomarker to predict the diseased condition.

Studies in mice models have indicated that the serum levels of three miRNAs, miR-21, miR-122, and miR-34a, were significantly upregulated in BALB/c, but not in C57BL/6 mice. This observation was correlated to the different degrees of hepatic necrosis in both the mice. Thus, it was hypothesized that the serum levels of these three circulating host miRNAs may reflect different hepatopathological outcomes owing to different immune responses induced in both mice after *S. japonicum* infection (Cai, Gobert et al. 2015, 2016). An ever-increasing amount of data regarding miRNA signatures associated with parasitic helminthiasis is being produced. Out of this, some are progressing towards translation to preliminary clinical applications. However, numerous questions regarding the complete biological functions of miRNAs and their role in the development of host pathology remain unanswered. It is also important to analyze the potential of circulating miRNAs as a novel diagnostic tool.

3.5   CONCLUDING REMARKS

There is growing evidence showing that miRNAs actively change the outcome of infections. Host miRNA dysregulation has been associated with impaired immune response and increased host colonization by the pathogen. Conversely, the host also employs miRNAs as a mechanism of defense against the parasite. Some studies have identified extracellular vesicles as intermediates in this communication, but in most cases it is not known precisely which molecules are involved. It appears that the parasites need to be viable to interfere with their host miRNome. Indeed, control cells exposed to heat-inactivated *Cryptosporidium* parasites displayed a miRNA expression profile similar to that of uninfected control samples (Winter, Gillan et al. 2015). Moreover, it has also been observed that the pattern of miRNA expression depends on post-infection timing and the multiplicity of infection. This suggests that the alteration of the host miRNome is not induced by a soluble factor or a membrane-exposed component such as pathogen-associated molecular patterns (PAMPs), but instead takes place via an active mechanism controlled by the invading parasite, raising questions about the nature and "modus operandi" of the parasite effectors that are presumably involved. In addition, similar studies would not only help us understanding the host–pathogen interaction, but would also help to dissect the biogenesis and function of miRNAs in normal, uninfected human cells. Furthermore, besides its potential as diagnostic and prognostic tools, miRNAs could also function as potential targets for chemo and immunotherapeutic strategies for parasitic diseases.

REFERENCES

Akira, S. and K. Takeda. 2004. Toll-like receptor signalling. *Nat Rev Immunol.* 4(7): 499–511.

Atayde, V. D., C. Tschudi and E. Ullu. 2011. The emerging world of small silencing RNAs in protozoan parasites. *Trends Parasitol.* 27(7): 321–327.

Barker, K. R., Z. Lu, H. Kim et al. 2017. miR-155 Modifies inflammation, endothelial activation and blood-brain barrier dysfunction in cerebral malaria. *Mol Med.* 23: 24–33.

Bartel, D. P. 2004. MicroRNAs: Genomics, biogenesis, mechanism, and function. *Cell* 116(2): 281–297.

Baum, J., A. T. Papenfuss, G. R. Mair et al. 2009. Molecular genetics and comparative genomics reveal RNAi is not functional in malaria parasites. *Nucleic Acids Res.* 37(11): 3788–3798.

Bayer-Santos, E., C. Aguilar-Bonavides, S. P. Rodrigues et al. 2013. Proteomic analysis of Trypanosoma cruzi secretome: Characterization of two populations of extracellular vesicles and soluble proteins. *J Proteome Res.* 12(2): 883–897.

Bayer-Santos, E., F. M. Lima, J. C. Ruiz et al. 2014. Characterization of the small RNA content of *Trypanosoma cruzi* extracellular vesicles. *Mol Biochem Parasitol.* 193(2): 71–74.

Bayer-Santos, E., M. M. Marini and J. F. da Silveira. 2017. Non-coding RNAs in host–pathogen interactions: Subversion of mammalian cell functions by protozoan parasites. *Front Microbiol.* 8: 474.

Bose, M., B. Barman, A. Goswami and S. N. Bhattacharyya. 2017. Spatiotemporal uncoupling of microRNA-mediated translational repression and target RNA degradation controls microRNP recycling in mammalian cells. *Mol Cell Biol.* 37(4): e00464–16.

Burke, M. L., L. McGarvey, H. J. McSorley et al. 2011. Migrating *Schistosoma japonicum* schistosomula induce an innate immune response and wound healing in the murine lung. *Mol Immunol.* 49(1–2): 191–200.

Burke, M. L., D. P. McManus, G. A. Ramm et al. 2010. Co-ordinated gene expression in the liver and spleen during Schistosoma japonicum infection regulates cell migration. *PLoS Negl Trop Dis.* 4(5): e686.

Cai, P., G. N. Gobert and D. P. McManus. 2016. MicroRNAs in parasitic helminthiases: Current status and future perspectives. *Trends Parasitol.* 32(1): 71–86.

Cai, P., G. N. Gobert, H. You et al. 2015. Circulating miRNAs: Potential novel biomarkers for hepatopathology progression and diagnosis of *Schistosomiasis japonica* in two murine models. *PLoS Negl Trop Dis.* 9(7): e0003965.

Cai, P., N. Hou, X. Piao et al. 2011. Profiles of small non-coding RNAs in *Schistosoma japonicum* during development. *PLoS Negl Trop Dis.* 5(8): e1256.

Cai, P., X. Piao, L. Hao et al. 2013. A deep analysis of the small non-coding RNA population in *Schistosoma japonicum* eggs. *PLoS One.* 8(5): e64003.

Cai, P., X. Piao, S. Liu et al. 2013. MicroRNA-gene expression network in murine liver during *Schistosoma japonicum* infection. *PLoS One.* 8(6): e67037.

Cai, Y., H. Chen, L. Jin et al. 2013. STAT3-dependent transactivation of miRNA genes following *Toxoplasma gondii* infection in macrophage. *Parasite Vect.* 6: 356.

Cestari, I., E. Ansa-Addo, P. Deolindo. 2012. *Trypanosoma cruzi* immune evasion mediated by host cell-derived microvesicles. *J Immunol.* 188(4): 1942–1952.

Chamnanchanunt, S., C. Kuroki, V. Desakorn et al. 2015. Downregulation of plasma miR-451 and miR-16 in *Plasmodium vivax* infection. *Exp Parasitol.* 155: 19–25.

Chen, X., Y. Ba, L. Ma et al. 2008. Characterization of microRNAs in serum: A novel class of biomarkers for diagnosis of cancer and other diseases. *Cell Res.* 18(10): 997–1006.

Chen, X. M., S. P. O'Hara, J. B. Nelson et al. 2005. Multiple TLRs are expressed in human cholangiocytes and mediate host epithelial defense responses to Cryptosporidium parvum via activation of NF-kappaB. *J Immunol.* 175(11): 7447–7456.

Chen, X. M., P. L. Splinter, S. P. O'Hara et al. 2007. A cellular micro-RNA, let-7i, regulates Toll-like receptor 4 expression and contributes to cholangiocyte immune responses against *Cryptosporidium parvum* infection. *J Biol Chem.* 282(39): 28929–28938.

Chhabra, R., Y. K. Adlakha, M. Hariharan et al. 2009. Upregulation of miR-23a-27a-24-2 cluster induces caspase-dependent and -independent apoptosis in human embryonic kidney cells. *PLoS One.* 4(6): e5848.

Costa, F. T., S. C. Lopes, M. Ferrer et al. 2011. On cytoadhesion of *Plasmodium vivax*: Raison d'etre? *Mem Inst Oswaldo Cruz.* 106(1): 79–84.

Coulson, R. M., N. Hall and C. A. Ouzounis. 2004. Comparative genomics of transcriptional control in the human malaria parasite *Plasmodium falciparum*. *Genome Res.* 14(8): 1548–1554.

Dalzell, J. J., P. McVeigh, N. D. Warnock et al. 2011. RNAi effector diversity in nematodes. *PLoS Negl Trop Dis.* 5(6): e1176.

DaRocha, W. D., K. Otsu, S. M. Teixeira et al. 2004. Tests of cytoplasmic RNA interference (RNAi) and construction of a tetracycline-inducible T7 promoter system in *Trypanosoma cruzi*. *Mol Biochem Parasitol*. 133(2): 175–186.

Delic, D., M. Dkhil, S. Al-Quraishy. 2011. Hepatic miRNA expression reprogrammed by *Plasmodium chabaudi* malaria. *Parasitol Res*. 108(5): 1111–1121.

El-Assaad, F., C. Hempel, V. Combes et al. 2011. Differential microRNA expression in experimental cerebral and noncerebral malaria. *Infect Immun*. 79(6): 2379–2384.

Eliaz, D., S. Kannan, H. Shaked et al. 2017. Exosome secretion affects social motility in *Trypanosoma brucei*. *PLoS Pathog*. 13(3): e1006245.

Fernandez-Calero, T., R. Garcia-Silva, A. Pena et al. 2015. Profiling of small RNA cargo of extracellular vesicles shed by *Trypanosoma cruzi* reveals a specific extracellular signature. *Mol Biochem Parasitol*. 199(1–2): 19–28.

Ferreira, L. R., A. F. Frade, R. H. Santos et al. 2014. MicroRNAs miR-1, miR-133a, miR-133b, miR-208a and miR-208b are dysregulated in chronic chagas disease cardiomyopathy. *Int J Cardiol*. 175(3): 409–417.

Fitzpatrick, J. M., E. Peak, S. Perally et al. 2009. Anti-schistosomal intervention targets identified by lifecycle transcriptomic analyses. *PLoS Negl Trop Dis*. 3(11): e543.

Frank, B., A. Marcu, A. L. de Oliveira Almeida Petersen et al. 2015. Autophagic digestion of Leishmania major by host macrophages is associated with differential expression of BNIP3, CTSE, and the miRNAs miR-101c, miR-129, and miR-210. *Parasite Vect*. 8: 404.

Garcia-Silva, M. R., R. F. das Neves, F. Cabrera-Cabrera et al. 2014. Extracellular vesicles shed by *Trypanosoma cruzi* are linked to small RNA pathways, life cycle regulation, and susceptibility to infection of mammalian cells. *Parasitol Res*. 113(1): 285–304.

Geraci, N. S., J. C. Tan and M. A. McDowell. 2015. Characterization of microRNA expression profiles in Leishmania-infected human phagocytes. *Parasite Immunol*. 37(1): 43–51.

Ghosh, J., M. Bose, S. Roy. 2013. Leishmania donovani targets Dicer1 to downregulate miR-122, lower serum cholesterol, and facilitate murine liver infection. *Cell Host Microbe*. 13(3): 277–288.

Gobert, G. N., L. Moertel, P. J. Brindley et al. 2009. Developmental gene expression profiles of the human pathogen *Schistosoma japonicum*. *BMC Genomics*. 10: 128.

Gomes, M. S., F. J. Cabral, L. K. Jannotti-Passos et al. 2009. Preliminary analysis of miRNA pathway in *Schistosoma mansoni*. *Parasitol Int*. 58(1): 61–68.

Gong, A. Y., G. Hu, R. Zhou et al. 2011. MicroRNA-221 controls expression of intercellular adhesion molecule-1 in epithelial cells in response to *Cryptosporidium parvum* infection. *Int J Parasitol*. 41(3–4): 397–403.

Hakimi, M. A. and D. Cannella. 2011. Apicomplexan parasites and subversion of the host cell microRNA pathway. *Trends Parasitol*. 27(11): 481–486.

Hall, N., M. Karras, J. D. Raine et al. 2005. A comprehensive survey of the Plasmodium life cycle by genomic, transcriptomic, and proteomic analyses. *Science*. 307(5706): 82–86.

Han, H., J. Peng, Y. Hong et al. 2013. MicroRNA expression profile in different tissues of BALB/c mice in the early phase of *Schistosoma japonicum* infection. *Mol Biochem Parasitol*. 188(1): 1–9.

He, X., X. Sai, C. Chen et al. 2013. Host serum miR-223 is a potential new biomarker for *Schistosoma japonicum* infection and the response to chemotherapy. *Parasite Vect*. 6: 272.

He, X., J. Xie, D. Zhang et al. 2015. Recombinant adeno-associated virus-mediated inhibition of microRNA-21 protects mice against the lethal schistosome infection by repressing both IL-13 and transforming growth factor beta 1 pathways. *Hepatology*. 61(6): 2008–2017.

Hentzschel, F., C. Hammerschmidt-Kamper, K. Borner et al. 2014. AAV8-mediated in vivo overexpression of miR-155 enhances the protective capacity of genetically attenuated malarial parasites. *Mol Ther*. 22(12): 2130–2141.

Hoogstrate, S. W., R. J. Volkers, M. G. Sterken et al. 2014. Nematode endogenous small RNA pathways. *Worm.* 3: e28234.

Hoy, A. M., R. J. Lundie, A. Ivens et al. 2014. Parasite-derived microRNAs in host serum as novel biomarkers of helminth infection. *PLoS Negl Trop Dis.* 8(2): e2701.

Hu, G., R. Zhou, J. Liu. 2010. MicroRNA-98 and let-7 regulate expression of suppressor of cytokine signaling 4 in biliary epithelial cells in response to Cryptosporidium parvum infection. *J Infect Dis.* 202(1): 125–135.

Hunt, N. H. and G. E. Grau. 2003. Cytokines: Accelerators and brakes in the pathogenesis of cerebral malaria. *Trends Immunol.* 24(9): 491–499.

Jia, B., Z. Chang, X. Wei et al. 2014. Plasma microRNAs are promising novel biomarkers for the early detection of *Toxoplasma gondii* infection. *Parasie Vect.* 7: 433.

Kosaka, N., H. Iguchi and T. Ochiya. 2010. Circulating microRNA in body fluid: A new potential biomarker for cancer diagnosis and prognosis. *Cancer Sci* 101(10): 2087–2092.

Lackner, P., C. Burger, K. Pfaller et al. 2007. Apoptosis in experimental cerebral malaria: Spatial profile of cleaved caspase-3 and ultrastructural alterations in different disease stages. *Neuropathol Appl Neurobiol* 33(5): 560–571.

Lambertz, U., M. E. Oviedo Ovando, E. J. Vasconcelos et al. 2015. Small RNAs derived from tRNAs and rRNAs are highly enriched in exosomes from both old and new world Leishmania providing evidence for conserved exosomal RNA packaging. *BMC Genomics* 16: 151.

LaMonte, G., N. Philip, J. Reardon et al. 2012. Translocation of sickle cell erythrocyte microRNAs into *Plasmodium falciparum* inhibits parasite translation and contributes to malaria resistance. *Cell Host Microbe* 12(2): 187–199.

Lemaire, J., G. Mkannez, F. Z. Guerfali et al. 2013. MicroRNA expression profile in human macrophages in response to Leishmania major infection. *PLoS Negl Trop Dis* 7(10): e2478.

Li, J. and L. S. Cox. 2000. Isolation and characterisation of a cAMP-dependent protein kinase catalytic subunit gene from *Plasmodium falciparum. Mol Biochem Parasitol* 109(2): 157–163.

Linhares-Lacerda, L. and A. Morrot. 2016. Role of small RNAs in trypanosomatid infections. *Front Microbiol.* 7: 367.

Mantel, P. Y., D. Hjelmqvist, M. Walch. 2016. Infected erythrocyte-derived extracellular vesicles alter vascular function via regulatory Ago2-miRNA complexes in malaria. *Nat Com.* 7: 12727.

Marie, C., H. P. Verkerke, D. Theodorescu et al. 2015. A whole-genome RNAi screen uncovers a novel role for human potassium channels in cell killing by the parasite *Entamoeba histolytica. Sci Rep.* 5: 13613.

Masaki, S., R. Ohtsuka, Y. Abe. 2007. Expression patterns of microRNAs 155 and 451 during normal human erythropoiesis. *Biochem Biophys Res Commun.* 364(3): 509–514.

Militello, K. T., P. Refour, C. A. Comeaux et al. 2008. Antisense RNA and RNAi in protozoan parasites: Working hard or hardly working? *Mol Biochem Parasitol.* 157(2): 117–126.

Mitchell, P. S., R. K. Parkin, E. M. Kroh et al. 2008. Circulating microRNAs as stable blood-based markers for cancer detection. *Proc Natl Acad Sci. U S A* 105(30): 10513–10518.

Navarro, I. C., F. M. Ferreira, H. I. Nakaya et al. 2015. MicroRNA transcriptome profiling in heart of *Trypanosoma cruzi*–infected mice: Parasitological and cardiological outcomes. *PLoS Negl Trop Dis.* 9(6): e0003828.

Nelson, P. T., M. De Planell-Saguer, S. Lamprinaki et al. 2007. A novel monoclonal antibody against human Argonaute proteins reveals unexpected characteristics of miRNAs in human blood cells. *RNA* 13(10): 1787–1792.

Nogueira, P. M., K. Ribeiro, A. C. Silveira et al. 2015. Vesicles from different Trypanosoma cruzi strains trigger differential innate and chronic immune responses. *J Extracell Vesicles* 4: 28734.

O'Hara, S. P., P. L. Splinter, G. B. Gajdos et al. 2010. NFkappaB p50-CCAAT/enhancer-binding protein beta (C/EBPbeta)-mediated transcriptional repression of microRNA let-7i following microbial infection. *J Biol Chem.* 285(1): 216–225.

Ono, T., L. Cabrita-Santos, R. Leitao et al. 2008. Adenylyl cyclase alpha and cAMP signaling mediate Plasmodium sporozoite apical regulated exocytosis and hepatocyte infection. *PLoS Pathog.* 4(2): e1000008.

Penet, M. F., A. Viola, S. Confort-Gouny et al. 2005. Imaging experimental cerebral malaria in vivo: Significant role of ischemic brain edema. *J Neurosci.* 25(32): 7352–7358.

Plattner, F. and D. Soldati-Favre 2008. Hijacking of host cellular functions by the Apicomplexa. *Annu Rev Microbiol.* 62: 471–487.

Ramirez, M. I., P. Deolindo, I. J. de Messias-Reason et al. 2017. Dynamic flux of microvesicles modulate parasite–host cell interaction of *Trypanosoma cruzi* in eukaryotic cells. *Cell Microbiol.* 19:e12672 10.1111/cmi.12672.

Rausch, M., M. Ecsedi, H. Bartake et al. 2015. A genetic interactome of the let-7 microRNA in *C. elegans. Dev Biol.* 401(2): 276–286.

Robinson, K. A. and S. M. Beverley. 2003. Improvements in transfection efficiency and tests of RNA interference (RNAi) approaches in the protozoan parasite Leishmania. *Mol Biochem Parasitol.* 128(2): 217–228.

Saba, R., D. L. Sorensen and S. A. Booth. 2014. MicroRNA-146a: A dominant, negative regulator of the innate immune response. *Front Immunol.* 5: 578.

Shapira, S., K. Speirs, A. Gerstein et al. 2002. Suppression of NF-kappaB activation by infection with *Toxoplasma gondii. J Infect Dis.* 185 Suppl 1: S66–S72.

Sheedy, F. J., E. Palsson-McDermott, E. J. Hennessy et al. 2010. Negative regulation of TLR4 via targeting of the proinflammatory tumor suppressor PDCD4 by the microRNA miR-21. *Nat Immunol.* 11(2): 141–147.

Singh, A. K., R. K. Pandey and R. Madhubala. 2016. MicroRNA expression profiling of *Leishmania donovani*–infected host cells uncovers the regulatory role of MIR30A-3p in host autophagy. *Autophagy.* 12(10): 1817–1831.

Starkey Lewis, P. J., M. Merz, P. Couttet et al. 2012. Serum microRNA biomarkers for drug-induced liver injury. *Clin Pharmacol Ther.* 92(3): 291–293.

Stoco, P. H., G. Wagner, C. Talavera-Lopez et al. 2014. Genome of the avirulent human-infective trypanosome—*Trypanosoma rangeli. PLoS Negl Trop Dis.* 8(9): e3176.

Sun, L., L. Zhi, S. Shakoor et al. 2016. microRNAs involved in the control of innate immunity in *Candida* infected *Caenorhabditis elegans. Sci Rep.* 6: 36036.

Szempruch, A. J., S. E. Sykes, R. Kieft et al. 2016. Extracellular vesicles from *Trypanosoma brucei* mediate virulence factor transfer and cause host anemia. *Cell* 164(1–2): 246–257.

Torrecilhas, A. C., R. I. Schumacher, M. J. Alves et al. 2012. Vesicles as carriers of virulence factors in parasitic protozoan diseases. *Microbes Infect.* 14(15): 1465–1474.

Tourneur, L. and G. Chiocchia 2010. FADD: A regulator of life and death. *Trends Immunol.* 31(7): 260–269.

Trager, W., G. S. Gill, C. Lawrence et al. 1999. Plasmodium falciparum: Enhanced gametocyte formation in vitro in reticulocyte-rich blood. *Exp Parasitol.* 91(2): 115–118.

Twu, O., N. de Miguel, G. Lustig et al. 2013. Trichomonas vaginalis exosomes deliver cargo to host cells and mediate hostratioparasite interactions. *PLoS Pathog.* 9(7): e1003482.

Wang, J. F., M. L. Yu, G. Yu et al. 2010. Serum miR-146a and miR-223 as potential new biomarkers for sepsis. *Biochem Biophys Res Commun.* 394(1): 184–188.

Winter, A. D., V. Gillan, K. Maitland et al. 2015. A novel member of the let-7 microRNA family is associated with developmental transitions in filarial nematode parasites. *BMC Genomics* 16: 331.

Xiao, C. and K. Rajewsky. 2009. MicroRNA control in the immune system: Basic principles. *Cell* 136(1): 26–36.

Xue, X., Q. Zhang, Y. Huang et al. 2008. No miRNA were found in *Plasmodium* and the ones identified in erythrocytes could not be correlated with infection. *Malaria J* 7: 47.

Zeiner, G. M., K. L. Norman, J. M. Thomson et al. 2010. *Toxoplasma gondii* infection specifically increases the levels of key host microRNAs. *PLoS One.* 5(1): e8742.

Zheng, Y., X. Cai and J. E. Bradley. 2013. microRNAs in parasites and parasite infection. *RNA Biol.* 10(3): 371–379.

Zhou, R., A. Y. Gong, A. N. Eischeid et al. 2012. miR-27b targets KSRP to coordinate TLR4-mediated epithelial defense against *Cryptosporidium parvum* infection. *PLoS Pathog.* 8(5): e1002702.

Zhou, R., G. Hu, J. Liu et al. 2009. NF-kappaB p65-dependent transactivation of miRNA genes following Cryptosporidium parvum infection stimulates epithelial cell immune responses. *PLoS Pathog.* 5(12): e1000681.

Zhu, D., X. He, Y. Duan et al. 2014. Expression of microRNA-454 in TGF-beta1-stimulated hepatic stellate cells and in mouse livers infected with *Schistosoma japonicum*. *Parasite Vectors* 7: 148.

4 Role of miRNA in Multiple Sclerosis

Krishnaraj Thirugnanasambantham,
Villianur Ibrahim Hairul Islam,
Subramanian Saravanan, Venugopal Senthil Kumar,
Ganapathy Ashok, and Muthiah Chellappandian

CONTENTS

Introduction

Multiple sclerosis (MS) is a chronic autoimmune neurodegenerative disorder of the central nervous system (CNS) characterized with inflammation and demyelination. When MS is present, the host immune system damages the protective layer of nerves (myelin) that leads to improper neural transmission. Accumulation of auto-reactive T cells and microglia/macrophages with MS has been reported to be correlated with axonal loss. MS is not a genetically inherited disease, but its development exists in certain genes. People with MS are expected to lead normal lives, and the clinical visibility of MS takes usually 20–40 years; however, its real onset appears much earlier (Ascherio and Munger 2007). Though MS is usually diagnosed in people 20 to 45 years of age, even very young children and older people have been reported with development of MS. The beginning symptoms of MS vary greatly, ranging from eye sight problems to slurred speech and can develop into paralysis or severe motor disabilities.

The clear etiology and pathogenesis of MS is still mysterious, but accumulating scientific evidence supports that demyelination of the axon in neurons occurs because of the migrating autoreactive lymphocytes across the blood-brain barrier (BBB) (Wu and Chen 2016). Though it is very challenging to pinpoint the direct reason behind

MS susceptibility, obtained epidemiological data assumes a significant role of child-hood events behind adult-onset of MS (Simpson et al. 2014). The immune system has been recognized as the major contributor to the susceptibility of MS (Oksenberg 2013), but until now it has been unclear how these immune-related genes in the central nervous system (CNS) were recognized as a specific target of inflamma-tion in MS. In general, four different patterns of MS, including relapsing-remitting multiple sclerosis (RRMS), primary progressive multiple sclerosis (PPMS), second-ary progressive multiple sclerosis (SPMS), and primary-relapsing multiple sclerosis (PRMS) have been suggested (Noseworthy et al. 2000).

Epigenetic events such as histone modifications, DNA methylation and non-coding RNA, are evidenced as factors regulating differentiation and function of immune cells (Qiu et al. 2017). The condition/tissue specific transcripts (coding and non-coding) of an organism are directly displayed from its transcriptomic datasets. MicroRNAs (miRNAs) are a family of endogenous, small, non-coding RNAs of length 19–24 nucleotides that modulate the expression of multiple target mRNAs at the posttranscriptional level via translational inhibition or mRNA degradation (Lund et al. 2004). They are essential in organized expression of genes involved in develop-ment, organogenesis, and homeostasis (Ebert and Sharp 2012). Accumulating evi-dence that has been generated from recent studies revealed that circulating miRNAs are involved in physiological and pathological processes and could be utilized as biomarkers, drug targets or therapeutic agents. Development of diseases such as immunological, neurological, cardiovascular disease and cancer are associated with dysregulated expression of miRNAs (Mendell et al. 2012). miRNAs are a fine regu-lator of inflammation (Saravanan et al. 2015). Apart from intracellular miRNAs, several body fluids, including plasma, serum, cerebrospinal fluid (CSF), breast milk, urine, tears, semen, and saliva were also reported as locations of miRNAs (Cortez et al. 2011).

Interestingly, researchers have reported a number of miRNA species differen-tially expressed in patients with MS compared with controls and these have the potential to be used as diagnostic biomarkers for drug-response. Overexpression of miRNA biogenesis machinery (Drosha, Dicer and DGCR8) may play a major role in MS pathogenesis (Jafari et al. 2015). The experimental autoimmune encephalo-myelitis (EAE), the mouse model of this chronic inflammatory and degenerative disease of the CNS, revealed dysregulated expression of miRNA has been linked to altered regulation in immune system and white matter (WM) pathology. MiR-142-3p is involved in inflammation-driven alterations of synaptic structure and its inhibition abolishes IL-1β- and glial glutamate-aspartate transporter-dependent synaptopathy in EAE wild-type mice (Mandolesi et al. 2017). Comparative microarray analysis of MS patient and healthy controls showed participation of hsa-miR-30a, hsa-miR-93, hsa-miR-20b, and hsa-miR-20a along with their respective mRNA targets in the pathophysiology of MS (Yang et al. 2017).

4.1 ROLE OF OLIGODENDROCYTE AND miRNA

Oligodendrocytes of mature CNS are involved in the production of a myelin sheath that provides support and insulation to axons. In MS the remyelination is mediated by proliferation, migration, and differentiation of oligodendroglial precursors cells (OPCs) into mature, myelin-producing oligodendrocytes at sites of demyelinated lesions. Maturation of oligodendrocytes to myelinating OPCs is a preplanned differentiation event finely refined by microRNAs, transcription factors and epigenetic factors. High levels of miR-125a-3p in the cerebrospinal fluid of MS patients with active demyelinating lesions and suggest its contribution in obstruction of OPC differentiation with impaired repair of demyelinated lesions (Lecca et al. 2016). Altering the miRNA biogenesis via Dicer deletion in oligodendroglia leads to demyelination, inflammation, and neurodegeneration (Zhao et al. 2010). This signifies the importance of miRNAs in myelin maintenance at later developmental stages. The study noted the importance both Dicer and miR-219 in oligodendrocyte differentiation and myelination (Dugas et al. 2010). Overexpression of lamin B1 was reported to be responsible for demyelination in CNS and miR-23 post-transcriptionally regulates lamin B1 transcript and enhances oligodendrocyte differentiation (Lin et al. 2014). The inflammatory cytokine tumor necrosis factor (TNF) exists both as transmembrane form (tmTNF) involved in TNF receptor 2 (TNFR2) and TNFR1 signaling and as soluble form (solTNF) involved in TNFR1 signaling (Figure 4.1). Recently it has been demonstrated that oligodendroglial TNFR2 as a key mediator of tmTNF-dependent protection in experimental autoimmune encephalomyelitis (EAE) and

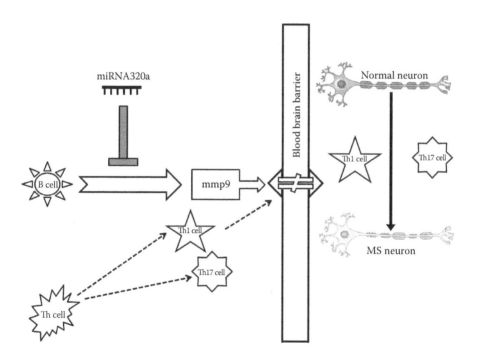

FIGURE 4.1 Role of miR-320a in regulating blood-brain barrier (BBB). The MMP9 produced by B cells are involved in disruption of BBB, which causes invasion of the Th cells and demyelization. Expression of miR-320a targets the MMP9 and prevents disruption of BBB, thereby preventing neurons from immune cell associated demyelization.

TNFR2 in the CNS have been reported as a drug target for the development of remyelinating event (Madsen et al. 2016). The abnormally high levels of miR-125a-3p in active demyelinating lesions of MS patients are responsible for affected OPC differentiation (Lecca et al. 2016). Recently, it was identified that among the number of miRNAs identified from mature OLs, miR-219 and miR-338 were identified to be abundantly expressed and evolutionarily conserved in vertebrate genomes. Wang et al. (2017a) demonstrated miR-219 that targets oligodendrocyte differentiation inhibitors including Lingo1 and Etv5 cooperates with miR-338 and enhances myelin restoration in EAE.

4.2 ROLE OF IMMUNE CELL IN MS PATHOGENESIS

Disruption of the blood-brain barrier (BBB) and activation of immune cells are central and early events of MS. Microglia are the main macrophages of CNS participating in phagocytosis of early demyelination and infiltration of blood-borne macrophages are responsible for axon debris clearance (Wu and Chen 2016). Upregulation of miR-200b in microglia showed significant decrease in iNOS expression, JNK pathways, pro-inflammatory cytokines, NO production and activated microglia migration; at the same time, inhibition of miR-200b reverses the microglia mediated neuro-inflammation (Jadhav et al. 2014). miRNAs downregulated in MS include miR-15a,b, miR-181c and miR-328 (Ma et al. 2014). Autoreactive T cells producing IFN-γ and IL-17 producing (T1 and T17 cells) are reported to mediate autoimmunity mediated axonal degeneration, demyelination in CNS ultimately leading to MS (Wu and Chen 2016). Regulatory T cells (Tregs cells) are other subpopulation of T cells that suppress the activation of the immune system, maintain homeostasis and provide tolerance to self-antigens. Though the molecular mechanism behind the suppressor/regulatory activity of Tregs has not been fully characterized, current research showed that Tregs cells require the contact with the cell being suppressed (Baecher-Allan et al. 2001; Viglietta et al. 2004). The balance between autoreactive T cells and Treg is crucial to the development of MS. Both extensive accumulation of macrophages and activation of microglia are common occurrences following neurological injury and are always engaged in active lesions (Wu and Chen 2016). In EAE, a mouse model of the human disease multiple sclerosis, the autoimmune inflammation of the CNS was characterized with microglia activation and infiltration of encephalitogenic T cells and leukocytes (Haanstra et al. 2013).

The B cells producing matrix metallopeptidase-9 (MMP-9) is involved in BBB disruption, and thus, maintains MS (Aung et al. 2015). The miR-320a targeting MMP-9 protein expression was reported to be decreased in B cells of MS patients (Figure 4.2) leading to disruption in BBB permeability and neurological disability (Aung et al. 2015).

Up-regulation of miR-17-5p and miR-193a were found in lymphocytes of relapsing–remitting MS patients (Lindberg et al. 2010). MiR-365 and miR-125b expression causes neuro-inflammation in amyotrophic lateral sclerosis (ALS) by interfering with IL-6 and STAT-3, respectively, and increasing TNF-α transcription (Parisi et al. 2013). The up-regulation of miR-16-2-3p has been observed in whole blood samples from RRMS treatment-naïve patients compared to controls (Keller et al. 2014).

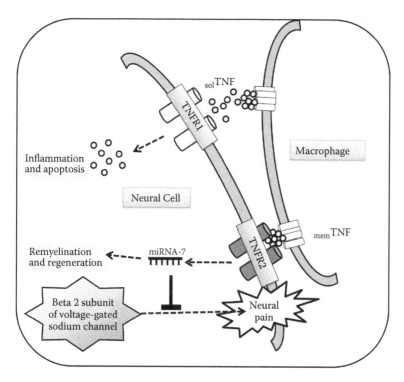

FIGURE 4.2 Differential role of membrane bound and soluble form of TNF signaling in neural cells. Expression of the soluble form of TNF activates neural inflammation/apoptosis by binding to and activating TNFR1 (TNF receptor 1). Activation of TNFR2 by soluble form of TNF, induces expression of miR-7 and leads to remyelization and regeneration. Expression of miR-7 attenuates neural pain via targeting beta 2 subunit of voltage-gated sodium channel.

4.2.1 ROLE OF TH CELL MiRNAS IN MS

Since multiple sclerosis has been considered as a T-cell mediated disorder, the pathogenic role of miRNAs in the dysregulation in T cell maturation, activation and function has been elucidated in detail (Table 4.1).

The miRNAs, including miR-21, miR-142-3p, miR-146a, miR-146b, miR-155 and miR-326, were reported to be up regulated in brain white matter lesions and PBMCs of MS patients and mouse models (Ma et al. 2014). MiR-155 is one of the first identified miRNA related to the T cells activation following TCR stimulation (Ksiazek-Winiarek et al. 2013). The upregulation of miR-155 in multiple immune cell lineages is connected to inflammation and multiple sclerosis by TLR ligands, inflammatory cytokines and specific antigens. The expression of miR-155 promotes pro-inflammatory functions of T-cells including Th-1 cell subsets and Th-17 (Thamilarasan et al. 2012). Inhibition of miR-155 showed normal lymphocyte development but altered Th1/Th2 ratio with increased Th2 polarization and Th2 cytokine production (Ksiazek-Winiarek et al. 2013). Recent studies showed that miR-155 plays a crucial role in inflammation in immune cells after TLR stimulation and contributes to neuro-inflammation in a TLR4-dependent manner in mice (Lippai et al. 2013).

TABLE 4.1

Tissue/Cell Specific Expression of MS Associated *miRNA* and Their Corresponding Targets

Sl.No	miRNA	Tissue/Cell	Host	Expression	Target	References
1	miR-142a-5p	Spinal cord	Mice with EAE	Upregulated	SOCS1	Talebi et al. 2017
2	miR-142a-3p	Spinal cord	Mice with EAE	Upregulated	TGFBR1	Talebi et al. 2017
3	miR-15b	Th17 cells	MS patients and mice with EAE	Downregulated	O-linked N-acetylglucosamine transferase	Liu et al. 2017
4	miR-142-3p	CSF	MS patients	Upregulated	Glial glutamate-aspartate transporter	Mandolesi et al. 2017
5	miR-27	T cells	Mice	Upregulated	c-Rel protein	Cruz et al. 2017
6	miR-328-3p	Whole blood sample	MS patients	Upregulated	RAC2	Yang et al. 2017
7	miR-125a-3p	DSF	MS patients	Upregulated	GPR17 and microtubule associated Protein MAP1B	Lecca et al. 2016
8	miR-223	CD4+T-cells	MS patients	Upregulated	FORKHEAD BOX O (FOXO1) and FOXO3	Hossein et al. 2016
9	miR-140-5p	PBMC	MS patients	Downregulated	STAT1	Guan et al. 2016
10	miR-30a	T cells	Mice with EAE	Downregulated	Traf3ip2	Wan et al. 2015; Qu et al. 2016
11	miR-146a	MS-active lesions	Mice with EAE	Upregulated	RhoA and nuclear factor of activated T cells 5	Wu et al. 2015
12	miR-320a	B cells	MS patients	Downregulated	Matrix metallopeptidase-9 (MMP-9)	Aung et al. 2015

Other miRNAs such as miR-146a and miR-21 have also been shown to be involved in modulation of immune response and Toll-like receptor (TLR) expression. These miRNAs are involved in neuro-inflammatory signaling pathways and mainly expressed in glial-derived cells including astrocytes (Brennan and Henshall 2017). Higher miR-27 levels in extracellular vesicles elevated from injured neurons transmigrates into adjacent microglia to activate neuro-inflammation (Sun et al. 2017). Interestingly, miRNAs let-7 and miR-21 have been recently identified as endogenous endosomal TLR7 agonists (Crews et al. 2017) and have been shown to induce neurodegeneration through neuronal TLR7 (Lehmann et al. 2012). Increased expression of miR-155 decreased the expression of neural stem cell (NSC) self-renewal genes by targeting the transcription factor C/EBPβ. Furthermore, targeting miR-155 with complementary anti-miRNA oligonucleotides significantly decreased the pathogenesis of neuro-inflammation and impairment in animal models of MS (Obora et al. 2017).

Dysregulation in miRNAs targeting SOCS6 was shown using a next-generation sequencing approach and downregulation of five miRNAs (miR-21-5p, miR-26b-5p, miR-29b-3p, miR-142-3p, and miR-155-5p) was confirmed using qPCR in CD4+ T cells of SPMS (Sanders et al. 2016). Nineteen differentially expressed miRNAs targeting TGFβ signaling pathway responsible for the regulatory T cell defect were observed in patients with multiple sclerosis (Severin et al. 2016). Over expression of miR-30a in naïve T cells inhibited its differentiation into Th17 and under *in vivo* condition it alleviated EAE by targeting interleukin-21 receptor (IL-21R) (Qu et al. 2016). The miR-223, a myeloid cell-specific miRNA was reported to be up regulated in MS and miR-223-knockout mice displayed reduced experimental autoimmune encephalomyelitis with reduced myeloid dendritic cells (mDCs) and Th17 cells in the CNS (Ifergan et al. 2016). miR-140-5p targeting STAT1 was downregulated in MS patients and synthetic miR-140-5p transfection reduced encephalitogenic Th1 differentiation (Guan et al. 2016). Genetic study revealed G>C variation in rs2910164 genotype of miR-146a plays certain role in susceptibility of MS in females and increased the release of proinflammatory cytokines TNF-α and IFN-γ without increasing the level of IL-1β (Li et al. 2015).

The expression of miR-146a was found to be up regulated in Th1 lymphocytes but downregulated in Th2 mouse lymphocytes cells (Monticelli et al. 2005). MiR-146 was upregulated upon TLR activation by LPS or by elevated levels of pro-inflammatory cytokines (TNF-α and IL-1β) during MS and EAE (Bergman et al. 2013). The Th1 polarization might also be regulated by miR-21, a direct potential target to IL-12p35 that controls Th1 cytokines IFN-γ secretion by the synergistic action with IL-18 (Ksiazek-Winiarek et al. 2013). Predominant expression of miR-128 in T cells results in a shift from Th2 to Th1/Th17 differentiation and cytokines production in susceptible DA rats after EAE induction (Guerau-de-Arellano et al. 2011). Apart from this, miR-15b targeting O-linked N-acetylglucosamine transferase that mediates O-linked N-acetylglucosamine glycosylation of NF-κB has been recognized as an important contributor involved in Th17 differentiation and the pathogenesis of MS (Liu et al. 2017). The increased levels of miR-29b were observed in memory CD4+ T cells isolated from multiple sclerosis (MS) patients, which may reflect Th1 mediated chronic inflammation by regulating T-bet and IFN-γ via a direct interaction with the 3′ UTR (Smith et al. 2012). The up regulation of miR-148a has been reported in the brain in inactive MS lesions and in Treg cells from MS patients versus controls (Junker, Hohlfeld, and Meinl 2011).

4.2.2 Role of Treg Cell miRNAs in MS

Mature forms of miR-142a were reported to be up regulated in spinal cords of EAE mice upon stimulation with either peptides of myelin oligodendrocyte glycoprotein (MOG) or anti-CD3/anti-CD28 antibodies (Table 4.1). Overexpression of miR-142a-5p targeting SOCS1 and TGFBR-1 transcripts was reported to influence T cell differentiation, thus, the pathogenesis of autoimmune neuroinflammation (Talebi et al. 2017). Overexpression of murine miR-27 in T cells severely impaired Treg differentiation with diminished homeostasis and immune suppressor function, which shows its role in Treg-mediated immunological tolerance in MS (Cruz et al. 2017). MiR-146a functions as an anti-inflammatory molecule by inhibiting NF-κB activity in brain endothelial cells and has been proposed as an endogenous inhibitor of NF-κB associated with decreased leukocyte adhesion during neuroinflammation (Wu et al. 2015). The miR-873 targeting TNFα-induced protein 3 that was induced by IL-17 promotes production of inflammatory cytokines and chemokines (IL-6, TNF-α, MIP-2, and MCP-1/5) and induces pathological processes of EAE mice by promoting NF-κB activation (Liu et al. 2014). Another example of a miRNA involved in the pathogenesis of MS is miR-223. Moreover, the miR-15a, miR-19a, miR-22, miR-210 and miR-223 were upregulated in brain white matter tissues, regulatory T cells (Tregs), plasma, blood cells and PBMCs of MS patients. The upregulation of these miRNAs in Tregs suggest the crucial role of Tregs in the pathogenesis of MS (Ma et al. 2014). MiR-223 highly upregulated in CD4+ CD25+ Treg cells and the downregulation of miR-223 might prevent macrophage hyper activation (Junker, Hohlfeld, and Meinl 2011). The upregulation of miR-106b-25 cluster (miR93, miR-106b) and two members of the mir-17-92 paralog cluster on chromosome 13 (miR-19a and miR-19b) were found in Treg cells from MS patients. MiR-106b and miR-25 are well known to modulate T regulatory cell differentiation and maturation by TGF-β signaling pathway, thus promoting MS development (De Santis et al. 2010). Deregulation of miR-106b-25 and miR-17-92 was found in B lymphocytes of MS patients. These two miRNA clusters have been shown to interact with B cell receptor, phosphatidyl-inositol-3-kinase (PI3K) and phosphatase and tensin homology (PTEN) signaling and, therefore, play a significant role in the immune-pathogenesis of MS (Sievers et al. 2012).

In MS patients the increased expression of miR-128 and miR-27b in naïve T cells and miR-340 in memory CD4+ T cells suggest that these miRNAs inhibit the Th2 pathway favoring the development of pro-inflammatory myelin-specific T cells that mediate CNS pathology. Due to significant clinical value in MS, these three miRNAs could serve as potential biomarkers and therapeutic targets for MS (Guerau-de-Arellano et al. 2012). The expression of miR-18b, mir-493 and miR-599 was significantly regulated in the PBMC of MS patients that were experiencing relapse compared to healthy controls (Thamilarasan et al. 2012). miRNAs are also known to have disease-promoting functions in MS. Research studies found altered expression of various miRNAs in peripheral blood leukocytes and brain tissue of

MS patients. In MS patients, the expression miR-18b and miR-599 was significantly associated with disease relapses and miR-96 with remissions (Otaegui et al. 2009). The dysregulation of miR-let-7g and miR-150 were confirmed in MS patients compared to controls. Regarding let-7g, it targets Toll-like receptor 4 (TLR4) expression which plays a key role in host immunity while miR-150 is mainly involved in the B cell maturation (Martinelli-Boneschi et al. 2012). The study in RRMS patients showed statistically robust differences in the upregulation of miR-18b, miR-599 and miR-493 in the relapse to control comparison. At the same time, miR-148a, miR-184, miR-193 and miR-96 were also observed as promising candidates in RRMS during remission to control comparison (Guerau-de-Arellano et al. 2012). The function of miR-497 was upregulated in CD4+ T cells and B cells, but downregulated in CD8+ T cells in MS patients as compared to the control subjects (Junker, Hohlfeld, and Meinl 2011).

Apart from the studies that investigated miRNAs from PBMC, T cells, B cells and brain astrocytes in MS risk, researchers have also focused on circulating miRNA involved in MS which could serve as a potential prognostic and diagnostic biomarker for MS (Table 4.1). The increased levels of plasma miRNA (miR-614, miR-572, miR-648, miR-1826, miR-422a and miR-22) were found in significant levels in MS individuals (Siegel et al. 2012). Circulating miR-92a-1, which is involved in cell cycle regulation and cell signaling, was significantly upregulated in RRMS. RRMS patients also showed increased levels of circulating miR let-7, which regulate stem cell differentiation and T cell activation, activate Toll-like receptor 7, and are linked to neurodegeneration (Gandhi et al. 2013). The expression of miR-let-7d significantly correlated with the levels of IL-1β in MS patients. The increased expression of miR-145 in MS patients showed its potential as a biomarker for the diagnosis of MS in blood, plasma and serum (Søndergaard et al. 2013).

4.3 ROLE OF VIRAL miRNA IN MS

Although the clear reason behind the development of MS has not yet been determined, viruses are assumed to be an environmental factor for MS susceptibility, but to date, no specific virus has been identified. Altered immune response to a virus, but not a specific virus, is responsible for the development of MS. The interaction between miRNA and virus pathogenesis is an emerging theory in neuro-immunology. The miR-155 has been identified as a potent regulator of inflammatory process in response to viral infection in the CNS. It has been reported that miR-155 influenced neuro-inflammation in EAE, the mouse model of MS (Murugaiyan et al. 2011) and viral-induced neuro-inflammation (Dickey et al. 2016). MiR-155 plays a vital role by regulating CD4+ and CD8+ T cell accumulation, T cell cytokine production (IFN-γ), NK cell maturation and expansion, astrogliosis and macrophage polarization in virally induced neuro-inflammation (Dickey et al. 2017).

Epstein-Barr virus (EBV) has been recognized as one of the etiologic factors for MS. Among the reported 44 miRNAs from MS patients, ebv-miR-BHRF1-2-5p and ebv-miR-BHRF1-3 were noticed to be significantly elevated in the circulation of MS patients and MALT1 (mucosa associated lymphoid tissue lymphoma transport protein 1), a key regulator of immune homeostasis was identified as a direct target of ebv-miR-BHRF1-2-5p (Wang et al. 2017b). Japanese macaque (JM) rhadinovirus (JMRV) isolated from inflammatory demyelinating encephalomyelitis (JME) showed clinical characteristics and immunohistopathology comparable to MS in humans. Screening of JMRV miRNAs involved in inflammatory signaling revealed that the viral miRNA miR-J8 has a seed sequence homologous to cellular miR-17/20/106 and miR-373 families involved in regulation of inflammatory process (Skalsky et al. 2012). The pro-inflammatory role of miR-29b was found during JEV induced neuro-inflammation and microglial activation via inhibition of anti-inflammatory protein, TNFAIP3, resulting in sustained activation of NF-κB, and subsequent secretion of pro-inflammatory cytokines. Thus, microglial activation appears to be regulated by miR-29b during neuro inflammation (Thounaojam et al. 2014).

4.4 EXPERIMENTAL AUTOIMMUNE ENCEPHALOMYELITIS (EAE) MODEL

The upregulation of selected immune-enriched miRNAs (miR-155, miR-142-3p, miR-146a/b, and miR-21) plays a pathogenic role in both human MS and experimental EAE lesions (Thounaojam et al. 2013). Increased expression of miR-142 isoforms (miR-142a-5p and miR-142a-3p) involved in the pathogenesis of autoimmune neuro-inflammation by interacting with SOCS1 and TGFBR-1 transcripts and influencing T cell differentiation have been reported in brain tissue from MS patient's brains and CNS tissue from EAE experimental mice (Talebi et al. 2017). Over-expression of miR-326 results in an increased number of Th17, while targeting miR-326 results in a decrease in the number of Th17 cells and severe symptoms of EAE (Thounaojam, Kaushik, and Basu 2013). Th-17 cell–associated miR-326 expression highly correlates with the pathogenesis of MS severity in patients and EAE in mice (Du et al. 2009). MiR-223 also regulates the number and function of myeloid-derived suppressor cells (MDSC) in MS and EAE. Modulating MDSC biology in EAE and in MS suggest the potential novel therapeutic applications of miR-223 (Cantoni et al. 2017). Th17-associated miR-15b was significantly downregulated in MS patients and EAE induced mice. Over-expression of miR-15b significantly alleviated EAE, whereas inhibition of this miRNA aggravated it (Liu et al. 2017). During the induction of EAE, the predominant expression of miR-181a in the lymphocyte population suggests its crucial role in T cell development and importance for TCR sensitivity (Bergman et al. 2013). Aryl hydrocarbon receptor (AHR) are helix-loop helix/Per-Arnt-Sim (bHLH/PAS) family of transcription factors that regulate interleukin-17 producing TH17 and Treg cells and play a critical role in autoimmune diseases including MS (Hanieh et al. 2016). Hanieh and Alzahrani (2013) suggested TCDD induced acetylcholinesterase-targeting miR-132 as a promising therapeutic target for anti-inflammatory treatment of EAE and MS. Upregulation of miR-132/212 cluster is observed during AHR activation. In addition, miR-212 targeted B-cell lymphoma 6 (a negative regulator of TH17 differentiation) and induced TH17 differentiation (Nakahama et al. 2013). Further, knockdown of the miR-132/212 cluster in mice correlated with higher resistance to EAE development.

4.5 miRNA ASSOCIATED WITH MS IN HUMAN SUBJECTS

In the environment of immunological disorders, activities of both innate and adaptive arms of the immune system are regulated by miRNAs. The critical role of miRNAs in major autoimmune diseases, including multiple sclerosis, tends to consider them as important pathogenic factor and/or potential therapeutic targets (Talebi et al. 2017). Increased levels of miR-146a in PBMCs from relapsing–remitting MS patients, suggest that miR-146a can be considered as a pro-inflammatory miRNA, and it could serve as a biomarker for MS disease progression (Cardoso, Guedes, and de Lima 2016). In RRMS patients, a significantly increased expression of miR-21 and miR-146a/b suggests a crucial role of these miRNAs in CD4+ cell activation in MS patients (Fenoglio et al. 2011). Further, the levels of miR-34a, miR-155, and miR-326 that target the 3′-UTR of CD47 were significantly elevated in active lesions of MS as compared to inactive lesions. Dysregulation of multiple miRNAs targeting CD47 regulatory protein such as miR-34a, miR-155, and miR-326 has been reported from the brains of MS patients (Junker et al. 2009). MiR-34a is upregulated in active MS lesions but downregulated in the blood samples of patients suffering from MS (Junker et al. 2011). Studies have shown that miRNA-326 is highly abundant in active MS lesions and is involved in the development of MS by promoting Th17 cell differentiation. The increased expression of miRNA-326 has also been reported in MS patients during relapses (Junker et al. 2011). T cell-specific deficiency of miR-17-92 targeting Ikaros Family Zinc Finger 4 (IKZF4) reduced TH17 differentiation and ameliorated EAE (Liu et al. 2014). The miRNAs, miR-155 and miR-326, that show dysregulated expression patterns could be used as diagnostic markers and therapeutic targets for MS (Ma et al. 2014). MiR-145, the most significant deregulated miRNA, allowed discriminating MS patients from healthy control groups with high specificity, sensitivity and accuracy, indicating miR-145 may also serve as a potential diagnostic biomarker for RRMS (Keller et al. 2009). MiR-145 and miR-20a-5p potentially mediate pleiotropic effects of IFN through MAPK pathway and these miRNAs have been reported as a potential diagnosis biomarker in treatment-naïve RRMS patients (Ehtesham et al. 2017).

Downregulation of miR-17 and miR-20a expression were found using Illumina Sentrix array matrix in naïve MS patients comprising primary progressive MS, secondary progressive MS, and relapsing–remitting MS (Cox et al. 2010). At the same time, decreased expression of miR-497, miR-1, and miR-126 were also found in lymphocytes of relapsing–remitting MS patients (Lindberg et al. 2010). Decreased expression of miR-20a-5p and hsa-miR-7-1-3p was also observed in whole blood samples from treatment-naïve patients with RRMS compared to healthy controls. Validating these miRNAs could be of significance in the pathophysiology and also promising biomarkers for MS (Keller et al. 2014). In circulating miRNAs signature, plasma miRNA (miR-1979) was significantly downregulated in MS individuals (Siegel et al. 2012). The significantly decreased miR-15b, miR-23a and miR-223 levels in MS patients could result in over-expression of target genes involved in disease pathogenesis of MS (Fenoglio et al. 2013).

4.6   TREATMENT STRATEGIES FOR MS AND ASSOCIATED miRNA

Multiple sclerosis is an autoimmune disorder leading to neurodegeneration and is one of the major health concerns around the world. MS patients are treated using immunomodulating drugs, which are monoclonal antibodies approved by the FDA, and they are administered to the patients either as injectable infusion or orally. MS patients are treated with humanized monoclonal antibody such as Natalizumab and Alemtuzumab and other effective drugs like Fingolimod, teriflunomide, Dimethyl fumarate, mitoxantrone, Glatiramer acetate, teriflunomide, Dimethyl fumarate and mitoxantrone (Table 4.2).

Natalizumab is commonly administered as a 300 mg IV infusion every 28 days, and the administration of this drug results in both upregulation and downregulation of many miRNA, including that of miR-18a, miR-20b, miR-29a, miR-103, and miR-326. Of these miR-326b is downregulated, whereas miR-18a, miR-20b, miR-29a, and miR-103 are upregulated in response to natalizumab treatment in MS patients (Ingwersen et al. 2015). MiR-20b and miR-326 play a significant role in regulating BBB breakdown and Th17 immune responses. In addition, miR-29a was shown specific and tightly regulated by modulating the differentiation of proinflammatory Th1 immune responses (Argaw et al. 2009; Smith et al. 2012).

Natalizumab also downregulates the expression of miR-17 and up regulates the levels of miR-106b (Petrocca, Vecchione, and Croce 2008). Both miR-106b and miR-17 have been observed as vital modulators of TGF- signaling and regulate the CD4+ T cell response by targeting TGFBR2 (Jiang et al. 2011). Natalizumab treatment in

TABLE 4.2
Drugs Associated with Expression of MS Related miRNA

Sl.No	Target miRNA	Drug/Treatment	Treatment Host	Expression Level	References
1	miR-15b, 23a and 223	Fingolimod	MS patients	Upregulated	Fenoglio et al. 2016
2	miR-125a-3p	Inhibition with an antago-miR	Cultured oligodendroglial precursors	Downregulated	Lecca et al. 2016
3	miR-155, miR-132	Natalizumab	MS patients	Downregulated	Mameli et al. 2016
4	miR-7a	TNFR2 ablation	Oligodendrocyte Precursor cells	Dysregulated	Madsen et al. 2016
5	miR-155	Dimethyl fumarate	MS patients	Downregulated	Michell-Robinson et al. 2016
6	miR-17	Natalizumab	MS patients	Downregulated	Meira et al. 2014
7	miR-16	3,3′-Diindolylmethane	Mice with EAE	Downregulated	Rouse et al. 2014

MS patient plays a vital role in the self-tolerance and regulation of T cell homeostasis by increasing the level of PTEN mRNA via downregulation of miR-17 expression (Newton and Turka 2012; Suzuki et al. 2001). IFNs are cytokines that show both anti-viral and anti-inflammatory response in MS patients and are believed to be regulated by a reduction of T-cell activation and IFN-γ production, altering of the BBB system and promotion of an anti-inflammatory immune response (Ransohoff and Cardona 2010). Interferon-beta therapy is commonly used in patients with MS (Hanieh and Alzahrani 2013). Interferon-beta treatment induced neuroprotection against the toxicity induced by activated microglia by suppressing the level of production of glutamate and superoxide by activated microglia (Jin et al. 2007; Yushchenko et al. 2003).

In MS patients, the expression level of miR-26a-5p significantly varied during different stages of interferon-based therapy. The levels of this miRNA were elevated in IFN-treated RRMS patients at 3 months' treatment and unchanged at 6 months' treatment in all MS patients. In contrast, continued treatment for 6 months decreased the level of postsynaptic density protein 95 (DLG4), which is involved in neuronal signalingvia, increasing the expression of miR-26a-5p (De Felice et al. 2014).

Moreover, around 20 miRNAs are reported to be involved in expression and play a significant role in the mechanisms of IFN therapy. Interestingly, seven of them, including let-7a-5p, let-7b-5p, miR-16-5p, miR-342-5p, miR-346, miR-518b and miR-760, were up regulated, whereas miR-27a-5p, miR-29a-3p, miR-29b-1-5p, miR-29c-3p, miR-95, miR-181c-3p, miR-193a-3p, miR-193-5p, miR-423-5p, miR-532-5p, miR-708-5p, miR-149-5p and miR-874, were downregulated in PBMCs from MS patients in response to IFN-treatment. In particular, some of the 20 miRNAs, such as the members of the miR-29 family, have been shown to be involved in apoptosis and in IFN signaling while others like hsa-miR-532-5p and hsa-miR-16-5p regulate IFN-responsive genes (Hecker et al. 2013). Further, miRNA can be used as a novel biomarker to predict and identify the effects of IFN-beta therapy for MS patients. Glatiramer acetate is an immunomodulation drug that downregulate miR-146a and miR-142-3p level in the PBMCs of MS patients. Glatiramer acetate mediated expansion of T regulatory cells expression is responsible for downregulation of miR-142-3p (Waschbisch et al. 2011). Downregulation of miR-146a in the glatiramer acetate treatment group may improve the disease status of MS patients by stimulating a Th1 to Th2 increase and inhibiting monocyte reactivity (Burger et al. 2009; Racke and Lovett-Racke 2011; Weber et al. 2004).

The GSK3 inhibitors were shown to be involved in increasing BBB tightness under physiologic conditions, decreasing inflammatory factors secreted in brain micro vascular endothelial cells and preventing BBB disruption by reducing monocyte adhesion to migration across the BBB system (Ramirez et al. 2010, 2013). GSK3 inhibitor may be involved in upregulation of miR-98 and let-7g* so as to attenuate leukocyte adhesion/migration into the BBB and diminished BBB permeability has been observed in both *in vitro* and *in vivo* models (Rom et al. 2015). Overexpression of miR-98 and let-7g* in brain endothelium contributes to inhibiting or suppressing the expression level of proinflammatory mediators, such as CCL2 and CCL5 (Rom et al. 2015).

The young and environmental enrichment (EE) serum-derived exosomes were reported to be enriched with miR-219 and nasal administration of such exomes to aging rats enhanced remyelination (Pusic et al. 2014). Anti-inflammatory effect of miR-124 "neurimmiR" has been identified in an animal model of multiple sclerosis. Peripheral administration of miR-124 mimics decreased neuro-inflammation by inhibiting the macrophages and myelin-specific T cells activation. Conversely, disease progression was observed in knockdown of miR-124 in microglia and macrophages that exacerbates further more cell activation (Cardoso et al. 2016). However, there is a conflict on this function, as miR-124 also elicited pro-inflammatory response and activate inflammation (Brennan and Henshall 2017). Over expression of miR-93 significantly reversed the expression of STAT3 and decreased neuropathic pain development and neuro-inflammation in chronic constriction injury (CCI) experimental rats. The results of this study suggesting that miR-93 may act as a potential therapeutic target for neuropathic pain intervention (Yan et al. 2017). Interestingly, miR-206 alleviated the development of neuropathic pain by directly targeting brain-derived neurotrophic factor (BDNF) and negatively mediating the MEK/ERK pathway (Sun et al. 2017). Exosomes of dendritic cells treated with low-level IFNγ (IFNγ-DC-Exos) that contain miRNAs responsible for remyelination have been suggested as the therapeutic agent for remyelination in multiple sclerosis (Pusic et al. 2014).

4.7 CONCLUSION

Noncoding RNAs, including miRNAs, are recorded as a major component in epigenetic regulation and are involved in refined control of neural gene expression. The accumulated literature on MS associated miRNAs reveal that CNS of MS patients are more susceptible to inflammation with reduced neuroprotective mechanisms. MS is an autoimmune disease and patients undergo several disease processes, including periphery inflammation, BBB damage, demyelination and CNS lesions. Association between circulatory miRNAs and magnetic resonance imaging (MRI) measures of disease severity revealed serum miRNAs as biomarkers for monitoring MS progression and suggested further validation (Regev et al. 2017). Identifying the role of these miRNAs will lead to controlling the expression of genes related to T cell activation and neuro inflammation in the pathogenesis of MS. Understanding the complex molecular mechanisms behind regulation of T cell, oligodendrocytes differentiation and myelin synthesis will result in unique insights into MS. Identifying targets of miRNA, and better understanding of RNA regulatory networks, will be helpful in further experiments for therapeutic agents against demyelinating diseases. To remove the bottlenecks in MS treatment strategies, it is necessary to unravel the specific miRNA-regulated checkpoints that control myelinogenesis/demyelation so that better diagnostic and/or therapeutic treatments can be devised for repairing the myelin in the CNS of MS patients.

REFERENCES

Argaw, A. T., Gurfein, B. T., Zhang, Y. et al. 2009. VEGF-mediated disruption of endothelial CLN-5 promotes blood-brain barrier breakdown. *Proceedings of the National Academy of Sciences of the United States of America* **106** (6): 1977–82.

Ascherio, A., and K. L. Munger. 2007. Environmental risk factors for multiple sclerosis. Part I: The role of infection. *Annals of Neurology* **61** (4): 288–99.

Aung, L. L., Mouradian, M. M., Dhib-Jalbut, S. et al. 2015. MMP-9 expression is increased in B lymphocytes during multiple sclerosis exacerbation and is regulated by microRNA-320a. *Journal of Neuroimmunology* **278**: 185–9.

Baecher-Allan, C., Brown, J. A., Freeman, G. J. et al. 2001. CD4+CD25 high regulatory cells in human peripheral blood. *Journal of Immunology (Baltimore, Md.: 1950)* **167** (3): 1245–53.

Bergman, P., James, T., Kular, L. et al. 2013. Next-generation sequencing identifies microRNAs that associate with pathogenic autoimmune neuroinflammation in rats. *Journal of Immunology (Baltimore, Md.: 1950)* **190** (8): 4066–75.

Brennan, G. P., and D. C. Henshall. 2017. microRNAs in the pathophysiology of epilepsy. *Neuroscience Letters* **S0304-3940** (17): 30027-7.

Burger, D., Molnarfi, N., Weber, M. S. et al. 2009. Glatiramer acetate increases IL-1 receptor antagonist but decreases T cell-induced IL-1beta in human monocytes and multiple sclerosis. *Proceedings of the National Academy of Sciences of the United States of America* **106** (11): 4355–9.

Cantoni, C., Cignarella, F., Ghezzi, L. et al. 2017. Mir-223 regulates the number and function of myeloid-derived suppressor cells in multiple sclerosis and experimental autoimmune encephalomyelitis. *Acta Neuropathologica* **133** (1): 61–77.

Cardoso, A. L., J. R. Guedes, and M. C. P. de Lima. 2016. Role of microRNAs in the regulation of innate immune cells under neuroinflammatory conditions. *Current Opinion in Pharmacology* **26**: 1–9.

Cortez, M. A., Bueso-Ramos, C., Ferdin, J. et al. 2011. MicroRNAs in body fluids—The mix of hormones and biomarkers. *Nature Reviews Clinical Oncology* **8** (8): 467–77.

Cox, M. B., Cairns, M. J., Gandhi, K. S. et al. 2010. MicroRNAs miR-17 and miR-20a inhibit T cell activation genes and are under-expressed in MS whole blood. *PloS One* **5** (8): e12132.

Crews, F. T., Walter, T. J., Coleman Jr, L. G. et al. 2017. Toll-like receptor signaling and stages of addiction. *Psychopharmacology (Berl)*. **234** (9–10): 1483–98.

Cruz, L. O., Hashemifar, S. S., Wu, C. et al. 2017. Excessive expression of miR-27 impairs Treg-mediated immunological tolerance. *The Journal of Clinical Investigation* **127** (2): 530–542.

De Felice, B., Mondola, P., Sasso, A. et al. 2014. Small non-coding RNA signature in multiple sclerosis patients after treatment with interferon-β. *BMC Medical Genomics* **7**: 26.

De Santis, G., Ferracin, M., Biondani, A. et al. (2010). Altered miRNA expression in T regulatory cells in course of multiple sclerosis. *Journal of Neuroimmunology* **226** (1–2): 165–71.

Dickey, L. L., Hanley, T. M., Huffaker, T. B. et al. 2017. MicroRNA 155 and viral-induced neuroinflammation. *Journal of Neuroimmunology* **S0165-5728** (16): 30466.

Dickey, L. L., Worne, C. L., Worne, J. L. et al. (2016). MicroRNA-155 enhances T cell trafficking and antiviral effector function in a model of coronavirus-induced neurologic disease. *Journal of Neuroinflammation* **13** (1): 240.

Du, C., Liu, C., Kang, J. et al. 2009. MicroRNA miR-326 regulates TH-17 differentiation and is associated with the pathogenesis of multiple sclerosis. *Nature Immunology* **10** (12): 1252–9.

Dugas, J. C., Cuellar, T. L., Scholze, A. et al. 2010. Dicer1 and miR-219 Are required for normal oligodendrocyte differentiation and myelination. *Neuron* **65** (5): 597–611.

Ebert, M. S., and P. A. Sharp. 2012. Roles for microRNAs in conferring robustness to biological processes. *Cell* **149** (3): 515–24.

Ehtesham, N., Khorvash, F. and M. Kheirollahi. 2017. miR-145 and miR20a-5p potentially mediate pleiotropic effects of interferon-beta through mitogen-activated protein kinase signaling pathway in multiple sclerosis patients. *Journal of Molecular Neuroscience* **61** (1): 16–24.

Fenoglio, C., Cantoni, C., De Riz, M. et al. 2011. Expression and genetic analysis of miRNAs involved in CD4+ cell activation in patients with multiple sclerosis. *Neuroscience Letters* **504** (1): 9–12.

Fenoglio, C., Ridolfi, E., Cantoni, C. et al. 2013. Decreased circulating miRNA levels in patients with primary progressive multiple sclerosis. *Multiple Sclerosis (Houndmills, Basingstoke, England)* **19** (14): 1938–42.

Fenoglio, C., De Riz, M., Pietroboni, A. M. et al. 2016. Effect of fingolimod treatment on circulating miR-15b, miR23a and miR-223 levels in patients with multiple sclerosis. *Journal of Neuroimmunology* **299**: 81–3.

Gandhi, R., Healy, B., Gholipour, T. et al. 2013. Circulating microRNAs as biomarkers for disease staging in multiple sclerosis. *Annals of Neurology* **73** (6): 729–40.

Guan, H., Singh, U. P., Rao, R. et al. 2016. Inverse correlation of expression of microRNA-140-5p with progression of multiple sclerosis and differentiation of encephalitogenic T helper type 1 cells. *Immunology* **147** (4): 488–98.

Guerau-de-Arellano, M., Alder, H., Ozer, H. G. et al. 2012. miRNA profiling for biomarker discovery in multiple sclerosis: From microarray to deep sequencing. *Journal of Neuroimmunology* **248** (1–2): 32–9.

Guerau-de-Arellano, M., Smith, K. M., Godlewski, J. et al. 2011. Micro-RNA dysregulation in multiple sclerosis favours pro-inflammatory T-cell-mediated autoimmunity. *Brain: A Journal of Neurology* **134** (Pt 12): 3578–89.

Haanstra, K. G., Wubben, J. A. M., Jonker, M. et al. 2013. Induction of encephalitis in rhesus monkeys infused with lymphocryptovirus-infected B-cells presenting MOG(34-56) peptide. *PloS One* **8** (8): e71549.

Hanieh, H., and A. Alzahrani. 2013. MicroRNA-132 suppresses autoimmune encephalomyelitis by inducing cholinergic anti-inflammation: a new Ahr-based exploration. *European Journal of Immunology* **43** (10): 2771–82.

Hanieh, H., Mohafez, O., Hairul-Islam, V. I. et al. 2016. Novel aryl hydrocarbon receptor agonist suppresses migration and invasion of breast cancer cells. *PloS One* **11**(12): e0167650.

Hecker, M., Thamilarasan, M., Koczan, D. et al. 2013. MicroRNA expression changes during interferon-beta treatment in the peripheral blood of multiple sclerosis patients. *International Journal of Molecular Sciences* **14** (8): 16087–110.

Hosseini, A., Ghaedi, K., Tanhaei, S. et al. 2016. Upregulation of CD4+ T-cell derived MiR-223 in the relapsing phase of multiple sclerosis patients. *Cell Journal (Yakhteh)* **18**(3): 371.

Ifergan, I., Chen, S., Zhang, B. et al. 2016. Cutting edge: MicroRNA-223 regulates myeloid dendritic cell-driven Th17 responses in experimental autoimmune encephalomyelitis. *Journal of Immunology (Baltimore, Md.: 1950)* **196** (4): 1455–9.

Ingwersen, J., Menge, T., Wingerath, B. et al. 2015. Natalizumab restores aberrant miRNA expression profile in multiple sclerosis and reveals a critical role for miR-20b. *Annals of Clinical and Translational Neurology* **2** (1): 43–55.

Jadhav, S. P., Kamath, S. P., Choolani, M. et al. 2014. microRNA-200b modulates microglia-mediated neuroinflammation via the cJun/MAPK pathway. *Journal of Neurochemistry* **130** (3): 388–401.

Jafari, N., Shaghaghi, H., Mahmoodi, D. et al. 2015. Overexpression of microRNA biogenesis machinery: Drosha, DGCR8 and Dicer in multiple sclerosis patients. *Journal of Clinical Neuroscience: Official Journal of the Neurosurgical Society of Australasia* **22** (1): 200–3.

Jiang, S., Li, C., Olive, V. et al. 2011. Molecular dissection of the miR-17-92 cluster's critical dual roles in promoting Th1 responses and preventing inducible Treg differentiation. *Blood* **118** (20): 5487–97.

Jin, S., Kawanokuchi, J., Mizuno, T. et al. 2007. Interferon-beta is neuroprotective against the toxicity induced by activated microglia. *Brain Research* **1179**: 140–6.

Junker, A. 2011. Pathophysiology of translational regulation by microRNAs in multiple sclerosis. *FEBS Letters* **585** (23): 3738–46.

Junker, A., R. Hohlfeld, and E. Meinl. 2011. The emerging role of microRNAs in multiple sclerosis. *Nature Reviews. Neurology* **7** (1): 56–9.

Junker, A., Krumbholz, M., Eisele, S. et al. 2009. MicroRNA profiling of multiple sclerosis lesions identifies modulators of the regulatory protein CD47. *Brain: A Journal of Neurology* **132** (Pt 12): 3342–52.

Keller, A., Leidinger, P., Lange, J. et al. 2009. Multiple sclerosis: microRNA expression profiles accurately differentiate patients with relapsing-remitting disease from healthy controls. *PloS One* **4** (10): e7440.

Keller, A., Leidinger, P., Steinmeyer, F. et al. 2014. Comprehensive analysis of microRNA profiles in multiple sclerosis including next-generation sequencing. *Multiple Sclerosis (Houndmills, Basingstoke, England)* **20** (3): 295–303.

Ksiazek-Winiarek, D. J., Kacperska, M. J., and A. Glabinski. 2013. MicroRNAs as novel regulators of neuroinflammation. *Mediators of Inflammation* **2013**: 172351.

Lecca, D., Marangon, D., Coppolino, G. T. et al. 2016. MiR-125a-3p timely inhibits oligodendroglial maturation and is pathologically up-regulated in human multiple sclerosis. *Scientific Reports* **6**: 34503.

Lehmann, S. M., Krüger, C., Park, B. et al. 2012. An unconventional role for miRNA: let-7 activates Toll-like receptor 7 and causes neurodegeneration. *Nature Neuroscience* **15** (6): 827–35.

Li, R., Rezk, A., Miyazaki, Y. et al. 2015. Proinflammatory GM-CSF-producing B cells in multiple sclerosis and B cell depletion therapy. *Science Translational Medicine* **7** (310): 310ra166.

Lin, S. T., Heng, M. Y., Ptáček, L. J. et al. 2014. Regulation of Myelination in the Central Nervous System by Nuclear Lamin B1 and Non-coding RNAs. *Translational Neurodegeneration* **3** (1): 4.

Lindberg, R. L. P., Hoffmann, F., Mehling, M. et al. 2010. Altered expression of miR-17-5p in CD4+ lymphocytes of relapsing-remitting multiple sclerosis patients. *European Journal of Immunology* **40** (3): 888–98.

Lippai, D., Bala, S., Csak, T. et al. 2013. Chronic alcohol-induced microRNA-155 contributes to neuroinflammation in a TLR4-dependent manner in mice. *PloS One* **8** (8): e70945.

Liu, R., Ma, X., Chen, L. et al. 2017. MicroRNA-15b Suppresses Th17 differentiation and is associated with pathogenesis of multiple sclerosis by targeting O-GlcNAc transferase. *Journal of Immunology (Baltimore, Md.: 1950)* **198** (7): 2626–2639.

Liu, Y., Carlsson, R., Comabella, M. et al. 2014. FoxA1 directs the lineage and immunosuppressive properties of a novel regulatory T cell population in EAE and MS. *Nature Medicine* **20** (3): 272–82.

Lund, J. M., Alexopoulou, L., Sato, A. et al. 2004. Recognition of single-stranded RNA viruses by Toll-like receptor 7. *Proceedings of the National Academy of Sciences of the United States of America* **101** (15): 5598–603.

Ma, X., Zhou, J., Zhong, Y. et al. 2014. Expression, regulation and function of microRNAs in multiple sclerosis. *International Journal of Medical Sciences* **11** (8): 810–8.

Madsen, P. M., Motti, D., Karmally, S. et al. 2016. Oligodendroglial TNFR2 mediates membrane TNF-dependent repair in experimental autoimmune encephalomyelitis by promoting oligodendrocyte differentiation and remyelination. *The Journal of Neuroscience: The Official Journal of the Society for Neuroscience* **36** (18): 5128–43.

Mameli, G., Arru, G., Caggiu, E. et al. 2016. Natalizumab therapy modulates miR-155, miR-26a and proinflammatory cytokine expression in MS patients. *PloS one* **11** (6): e0157153.

Mandolesi, G., De Vito, F., Musella, A. et al. 2017. miR-142-3p Is a key regulator of IL-1β-dependent synaptopathy in neuroinflammation. *The Journal of Neuroscience: The Official Journal of the Society for Neuroscience* **37** (3): 546–61.

Martinelli-Boneschi, F., Fenoglio, C., Brambilla, P. et al. 2012. MicroRNA and mRNA expression profile screening in multiple sclerosis patients to unravel novel pathogenic steps and identify potential biomarkers. *Neuroscience Letters* **508** (1): 4–8.

Meira, M., Sievers, C., Hoffmann, F. et al. 2014. Unraveling Natalizumab Effects on Deregulated miR-17 Expression in CD4. *Journal of Immunology Research 897249.* doi:10.1155/2014/897249

Mendell, J. R., Shilling, C., Leslie, N. D. et al. 2012. Evidence-based path to newborn screening for Duchenne muscular dystrophy. *Annals of Neurology* **71** (3): 304–13.

Michell-Robinson, M. A., Moore, C. S., Healy, L. M. et al. 2016. Effects of fumarates on circulating and CNS myeloid cells in multiple sclerosis. *Annals of Clinical and Translational Neurology* **3** (1): 27–41.

Monticelli, S., Ansel, K. M., Xiao, C. et al. 2005. MicroRNA profiling of the murine hematopoietic system. *Genome Biology* **6** (8): R71.

Murugaiyan, G., Beynon, V., Mittal, A. et al. 2011. Silencing microRNA-155 ameliorates experimental autoimmune encephalomyelitis. *Journal of Immunology (Baltimore, Md.: 1950)* **187** (5): 2213–21.

Nakahama, T., Hanieh, H., Nguyen, N. T. et al. 2013. Aryl hydrocarbon receptor-mediated induction of the microRNA-132/212 cluster promotes interleukin-17-producing T-helper cell differentiation. *Proceedings of the National Academy of Sciences of the United States of America* **110** (29): 11964–9.

Newton, R. H., and L. A. Turka. 2012. Regulation of T cell homeostasis and responses by pten. *Frontiers in Immunology* **3**: 151.

Noseworthy, J. H., Lucchinetti, C., Rodriguez, M. et al. 2000. Multiple sclerosis. *The New England Journal of Medicine* **343** (13): 938–52.

Obora, K., Onodera, Y., Takehara, T. et al. 2017. Inflammation-induced miRNA-155 inhibits self-renewal of neural stem cells via suppression of CCAAT/enhancer binding protein β (C/EBPβ) expression. *Scientific Reports* **7**: 43604.

Oksenberg, J. R. 2013. Decoding multiple sclerosis: An update on genomics and future directions. *Expert Review of Neurotherapeutics* **13** (12 Suppl): 11–9.

Otaegui, D., Baranzini, S. E., Armananzas, R. et al. 2009. Differential micro RNA expression in PBMC from multiple sclerosis patients. *PloS One* **4** (7): e6309.

Parisi, C., Arisi, I., Ambrosi, N. D. et al. 2013. Dysregulated microRNAs in amyotrophic lateral sclerosis microglia modulate genes linked to neuroinflammation. *Cell Death & Disease* **4**: e959.

Petrocca, F., A. Vecchione, and C. M. Croce. 2008. Emerging role of miR-106b-25/miR-17-92 clusters in the control of transforming growth factor beta signaling. *Cancer Research* **68** (20): 8191–4.

Pusic, A. D., Pusic, K. M., Clayton, B. L. L. et al. 2014. IFNγ-stimulated dendritic cell exosomes as a potential therapeutic for remyelination. *Journal of Neuroimmunology* **266** (1–2): 12–23.

Qiu, H., Wu, H., Chan, V. et al. 2017. Transcriptional and epigenetic regulation of follicular T-helper cells and their role in autoimmunity. *Autoimmunity* **50** (2): 71–81.

Qu, X., Zhou, J., Wang, T. et al. 2016. MiR-30a inhibits Th17 differentiation and demyelination of EAE mice by targeting the IL-21R. *Brain, Behavior, and Immunity* **57**: 193–9.

Racke, M. K., and A. Lovett-Racke. 2011. Glatiramer acetate treatment of multiple sclerosis: An immunological perspective. *Journal of Immunology (Baltimore, Md.: 1950)* **186** (4): 1887–90.

Ramirez, S. H., Fan, S., Dykstra, H. et al. 2013. Inhibition of glycogen synthase kinase 3β promotes tight junction stability in brain endothelial cells by half-life extension of occludin and claudin-5. *PloS One* **8** (2): e55972.

Ramirez, S. H., Fan, S., Zang, M. et al. 2010. Inhibition of glycogen synthase kinase 3beta (GSK3beta) decreases inflammatory responses in brain endothelial cells. *The American Journal of Pathology* **176** (2), 881–92.

Ransohoff, R. M., and A. E. Cardona. 2010. The myeloid cells of the central nervous system parenchyma. *Nature* **468** (7321): 253–62.

Regev, K., Healy, B. C., Khalid, F. et al. 2017. Association between serum microRNAs and magnetic resonance imaging measures of multiple sclerosis severity. *JAMA Neurology* **74** (3): 275–285.

Rom, S., Dykstra, H., Zuluaga-Ramirez, V. et al. 2015. miR-98 and let-7g* protect the blood-brain barrier under neuroinflammatory conditions. *Journal of Cerebral Blood Flow and Metabolism: Official Journal of the International Society of Cerebral Blood Flow and Metabolism* **35** (12): 1957–65.

Rouse, M., Rao, R., Nagarkatti, M. et al. 2014. 3, 3'-diindolylmethane ameliorates experimental autoimmune encephalomyelitis by promoting cell cycle arrest and apoptosis in activated T cells through microRNA signaling pathways. *Journal of Pharmacology and Experimental Therapeutics* **350** (2): 341–52.

Sanders, K. A., Benton, M. C., Lea, R. A. et al. 2016. Next-generation sequencing reveals broad down-regulation of microRNAs in secondary progressive multiple sclerosis CD4+ T cells. *Clinical Epigenetics* **8** (1): 87.

Saravanan, S., Thirugnanasambantham, K., Hanieh, H. et al. 2015. miRNA-24 and miRNA-466i-5p controls inflammation in rat hepatocytes. *Cellular & Molecular Immunology* **12** (1): 113–5.

Severin, M. E., Lee, P. W., Liu, Y. et al. 2016. MicroRNAs targeting TGFβ signalling underlie the regulatory T cell defect in multiple sclerosis. *Brain: A Journal of Neurology* **139** (Pt 6): 1747–61.

Siegel, S. R., Mackenzie, J., Chaplin, G. et al. 2012. Circulating microRNAs involved in multiple sclerosis. *Molecular Biology Reports* **39** (5): 6219–25.

Sievers, C., Meira, M., Hoffmann, F. et al. 2012. Altered microRNA expression in B lymphocytes in multiple sclerosis: Towards a better understanding of treatment effects. *Clinical Immunology (Orlando, Fla.)* **144** (1): 70–9.

Simpson, R., Booth, J., Lawrence, M. et al. 2014. Mindfulness based interventions in multiple sclerosis—A systematic review. *BMC Neurology* **14**: 15.

Skalsky, R. L., Corcoran, D. L., Gottwein, E. et al. 2012. The viral and cellular microRNA targetome in lymphoblastoid cell lines. *PLoS Pathogens* **8** (1): e1002484.

Smith, K. M., Guerau-de-Arellano, M., Costinean, S. et al. 2012. miR-29ab1 deficiency identifies a negative feedback loop controlling Th1 bias that is dysregulated in multiple sclerosis. *Journal of Immunology (Baltimore, Md.: 1950)* **189** (4): 1567–76.

Søndergaard, H. B., Hesse, D., Krakauer, M. et al. 2013. Differential microRNA expression in blood in multiple sclerosis. *Multiple Sclerosis (Houndmills, Basingstoke, England)*, **19** (14): 1849–57.

Sun, W., Zhang, M., Wang, Y. et al. 2017. Neuroprotective effects of miR-27a against traumatic brain injury via suppressing FoxO3a-mediated neuronal autophagy. *Biochemical and Biophysical Research Communications* **482** (4): 1141–47.

Suzuki, A., Yamaguchi, M. T., Ohteki, T. et al. 2001. T cell-specific loss of Pten leads to defects in central and peripheral tolerance. *Immunity* **14** (5): 523–34.

Talebi, F., Ghorbani, S., Chan, W. F. et al. 2017. MicroRNA-142 regulates inflammation and T cell differentiation in an animal model of multiple sclerosis. *Journal of Neuroinflammation* **14** (1): 55.

Thamilarasan, M., Koczan, D., Hecker, M. et al. 2012. MicroRNAs in multiple sclerosis and experimental autoimmune encephalomyelitis. *Autoimmunity Reviews* **11** (3): 174–9.

Thounaojam, M. C., Kaushik, D. K., and A. Basu. 2013. MicroRNAs in the brain: It's regulatory role in neuroinflammation. *Molecular Neurobiology* **47** (3): 1034–44.

Thounaojam, M. C., Kaushik, D. K., Kundu, K. et al. 2014. MicroRNA-29b modulates Japanese encephalitis virus-induced microglia activation by targeting tumor necrosis factor alpha-induced protein 3. *Journal of Neurochemistry* **129** (1): 143–54.

Viglietta, V., Baecher-Allan, C., Weiner, H. L. et al. 2004. Loss of functional suppression by CD4+CD25+ regulatory T cells in patients with multiple sclerosis. *The Journal of Experimental Medicine* **199** (7): 971–9.

Wan, Q., Zhou, Z., Ding, S. et al. 2015. The miR-30a negatively regulates IL-17-mediated signal transduction by targeting Traf3ip2. *Journal of Interferon & Cytokine Research* **35** (11): 917–23.

Wang, H., Moyano, A. L., Ma, Z. et al. 2017. miR-219 Cooperates with miR-338 in myelination and promotes myelin repair in the CNS. *Developmental Cell* **40** (6): 566–582.e5.

Wang, Y. F., He, D. D., Liang, H. W., 2017. The identification of up-regulated ebv-miR-BHRF1-2-5p targeting MALT1 and ebv-miR-BHRF1-3 in the circulation of patients with multiple sclerosis. *Clinical and Experimental Immunology* **189** (1): 120–6.

Waschbisch, A., Atiya, M., Linker, R. A. et al. 2011. Glatiramer acetate treatment normalizes deregulated microRNA expression in relapsing remitting multiple sclerosis. *PloS One* **6** (9): e24604.

Weber, M. S., Starck, M., Wagenpfeil, S. et al. 2004. Multiple sclerosis: glatiramer acetate inhibits monocyte reactivity in vitro and in vivo. *Brain: A Journal of Neurology* **127** (Pt 6): 1370–8.

Wu, D., Cerutti, C., Lopez-Ramirez, M. A. et al. 2015. Brain endothelial miR-146a negatively modulates T-cell adhesion through repressing multiple targets to inhibit NF-κB activation. *Journal of Cerebral Blood Flow and Metabolism* **35** (3): 412–23.

Wu, T., and G. Chen. 2016. miRNAs Participate in MS pathological processes and its therapeutic response. *Mediators of Inflammation* **2016**: 4578230.

Yan, X. T., Ji, L. J., Wang, Z. et al. 2017. MicroRNA-93 alleviates neuropathic pain through targeting signal transducer and activator of transcription 3. *International Immunopharmacology* **46**: 156–62.

Yang, Q., Pan, W. and L. Qian. 2017. Identification of the miRNA-mRNA regulatory network in multiple sclerosis. *Neurological Research* **39** (2): 142–51.

Yushchenko, M., Mader, M., Elitok, E. et al. 2003. Interferon-beta-1 b decreased matrix metalloproteinase-9 serum levels in primary progressive multiple sclerosis. *Journal of Neurology* **250** (10): 1224–8.

Zhao, X., He, X., Han, X. et al. 2010. MicroRNA-mediated control of oligodendrocyte differentiation. *Neuron* **65** (5): 612–26.

5 miRNA Dysregulation in Inflammatory Bowel Disease and Its Consequences

Jaishree Paul and Swati Valmiki

CONTENTS

Introduction

Inflammatory bowel disease (IBD) is an inflammatory disorder of the gastrointestinal tract. It is chronic, idiopathic and relapsing in nature and exhibits two clinical forms: i) Ulcerative Colitis (UC) and ii) Crohn's Disease (CD). The onset of IBD is known to prevail during the second and third decade of life and in a majority of affected individuals it shows progression towards chronic and relapsing disorder (Xavier and Podolsky 2007). The etiology of IBD has not been defined yet, but it is hypothesized that interplay of several factors such as microbial, environmental and genetic factors, contribute to the pathogenesis of this disease. The incidence rate is higher in the western world, with an increasing incidence of early onset of the disease. Of late, a substantial increase in the incidence rate in a different age group of individuals has been observed in many developing countries, mounting a heavy healthcare cost. The two forms of IBD, ulcerative colitis and Crohn's disease, differ on the basis of endoscopic, clinical and pathological features.

5.1 ULCERATIVE COLITIS

In UC, the mucosal inflammation initiates from the rectum and progresses upwards in a continuous manner, affecting the gastrointestinal tract to a varying degree. Besides severe mucosal inflammation, it includes production of proinflammatory mediators and development of superficial ulcers. UC manifests as thinning of the bowel wall and loss of vascular patterns. Histopathological studies reveal a massive infiltration of neutrophils in the lamina propria and crypts, where they eventually form micro abscesses. UC is characterized as a Th2 mediated disease with increased production of proinflammatory cytokines like Il-5 and IL-13. It has also been observed that people with UC have a high risk of developing colorectal cancer after eight to ten years of diagnosis.

5.2 CROHN'S DISEASE

Unlike UC, CD shows a discontinuous pattern of inflammation which is transmural in nature, affecting even the deeper layers of the intestinal tract. CD involves thickening of bowel wall, giving it a cobblestone appearance due to deep ulcers and swelling of tissues. The characteristic features observed in CD patients are macrophage aggregates, forming non-caseating granulomas which could involve any part of gastrointestinal tract, but the terminal ileum is known to get affected commonly. In CD, very early mucosal lesions appear in Peyer's patches. CD is mainly associated with abnormal Th1 response, which is induced from higher IL-18 and Il-2 levels in the intestinal mucosa (Geremia et al. 2014).

5.3 CONTRIBUTING FACTORS IN THE PATHOGENESIS OF IBD

5.3.1 MICROBIAL FACTORS

Dysbiosis in the gut microbiota is one of the major causes in disease pathogenesis. IBD patients manifest dysbiosis in their luminal microflora characterized by reduced microbial diversity as compared to healthy people. Certain bacterial phenotypes have been associated with disease pathogenesis, such as adherent invasive *Escherichia coli* (AIEC). Michaud et al. observed AIEC localized in the ileum of 22% of CD patients in comparison with 6.2% of controls. AIEC disrupts the epithelium and persists within the macrophages, indicating its probable role in pathogenesis of Crohn's disease (Darfeuille-Michaud et al. 2004). In addition, *Faecalibacterium prausnitzii*, a butyrate producing bacteria belonging to the Firmicutes, occurs more frequently in controls than IBD patients (Ananthakrishnan 2015).

5.3.2 Environmental Factors

Factors including diet, smoking, appendectomy, and hygiene contribute greatly to the disease pathogenesis (Darfeuille et al. 2004; Ananthakrishnan 2015).

5.3.3 Genetic Factors

Genetic factors include the various susceptibility genes associated with disease pathogenesis. Genome-wide association studies have successfully and efficiently identified some genetic loci for UC as well as CD. Some of these genetic loci are commonly associated with UC and CD while others are specific to the two disease forms. Genes such as *NOD1* (nucleotide-binding oligomerization domain protein 2, also known as CARD15), *ATG16L1* (encoding autophagy related 16-like protein 1) and *IRGM* (which encodes immunity-related GTPase family, M) are specifically associated with CD and not seen in patients with UC (Ananthakrishnan 2015). In contrast to this, genes involved in IL-23 pathway such as *IL-23R* (which encodes for the IL-23 receptor subunit), IL12B (which encodes the p40 subunit of IL-12 and IL-23) and STAT3 (encoding signal transducer and activator of transcription 3) are known to be associated with both UC and CD. Similar to this, NKX2-3 (NK2 transcription factor related, locus 3) which is a homeodomain-containing transcription factor also shows association with both the disease subtypes (Cho 2008).

5.3.4 Epigenetic Factors

Besides the crucial role of susceptibility genes in disease pathogenesis, epigenetic factors have emerged as critical regulators of inflammation during UC. The major epigenetic mechanisms are histone modification, DNA methylation, RNA interference and nucleosome positioning. DNA methylation is the most well-studied modification in context to IBD and reported to differ significantly in the inflamed versus non-inflamed colonic mucosa of UC and CD patients. Even the peripheral blood DNA methylation status of UC and CD patients differs from that of healthy individuals (Loddo and Romano 2015). Differential methylation of various genes has been reported in context to UC such as estrogen receptor gene, MLH1 (mutator L homolog 1), HPP1 (hyperplastic polyposis), PAR2 (protease activated receptor 2) and MDR1 (multidrug resistance gene 1). In addition, a genome wide DNA methylation profiling study of ileal CD patients reported differential DNA methylation in genes responsible for immune response and susceptibility genes of CD (Scarpa and Stylianou 2012).

5.4 MICRORNA AS EPIGENETIC REGULATOR IN IBD

MicroRNA have gained focus in the past few years due to their role in modulation of inflammatory response in various inflammatory and autoimmune disorders like rheumatoid arthritis, inflammatory bowel disease, psoriasis, multiple sclerosis and various cancers. The first report came from Wu et al. (2008) showing differentially expressed microRNAs in mucosa samples of UC patients at different stages of the disease. Since then differential expression profiles of microRNAs have been reported in both UC and CD patients, and they are implicated to modulate the inflammatory response by targeting the signaling molecules of major inflammatory pathways such as NFkB and MAPK and other crucial molecules such as tight junction proteins, which maintain the membrane integrity. Currently, various studies indicate evidence that the miRNAs add to the complexity of this multifactorial disease. Over 1800 miRNAs have now been identified that are functionally active, and each miRNA possesses the ability to target and modulate the expression of various gene transcripts and these miRNAs belong to the complex gene regulatory networks (GRNs) (Kozomara and Griffiths-Jones 2011).

5.4.1 DYSREGULATION OF MiRNA IN VARIOUS TISSUES OF MUCOSA

Intestinal myofibroblasts (IMF) derived from mesenchymal cells are actively involved in the inflammation process in the mucosa. Interestingly, miR-155 is known to be significantly higher in colonic mucosal IMF cells of UC patients but not in CD patients. This led to reduction in the expression level of its target gene SOCS1, due to which overexpression of cytokines ensue. Therefore, inhibiting the level of miR-155 could be an alternative therapeutic target for maintaining the cytokine homoeostasis during disease condition (Pathak et al. 2015). Also, elevated expression of miR-155 has been reported in pediatric CD patients where it targets the anti-inflammatory molecule Transforming Growth factor (TGFβ) present in the duodenal epithelial cells (Szucs et al. 2016). Cheng et al. (2015) showed down regulation of miR-19b in CD patients and simultaneous upregulation of SOCS3 (suppressor of cytokine signaling 3) responsible for chemokine production in intestinal epithelial cells. Further, the therapeutic potential of this miRNA was investigated. Inflammation could be controlled by introducing pre-miR-19b into intestinal epithelial cells in TNBS-induced colitis model. However, due to the complexity of the disease, designing miRNAs as biomarkers needs more research while addressing different sub-phenotypes of the disease encountered in IBD (Cheng et al. 2008).

5.4.2 ROLE IN INFLAMMATORY PATHWAYS

MicroRNA plays an important role in cell differentiation and activation, cell signaling and in regulating innate and adaptive immune response which are critical in regulating inflammatory responses. Dysregulation in these responses results in inappropriate and exaggerated inflammation which is a major cause of many acute and chronic inflammatory disorders like IBD. Innate immune response, which is the first line of defense against pathogens, primarily involves early recognition of antigens by pathogen associated molecular patterns (PAMPs). PAMPs then stimulate extracellular receptors called Toll like receptors (TLRs) and intracytoplasmic located nucleotide binding oligomerization domain containing proteins (NOD) like receptors to activate downstream signaling cascade via NFkB, mitogen activated kinase pathway (MAPK) and interferon regulatory factors. All these pathways are actively regulated by microRNAs. Human monocytes, upon exposure to TLR ligands like LPS, peptidoglycan and flagellin, induce miR-146a and 146b (Taganov et al. 2006). In the mice model, miR-146a is reported to interfere in the TLR signaling cascade by reducing the expression of two components of its pathway, TRAF6 and IRAK-1. This initiates a negative feedback loop, resulting in attenuated TLR response (Yang et al. 2012). Upregulation of miR-126 expression during active disease of UC led to downregulation of NF-KB inhibitor IKBα expression both at mRNA and protein level, resulting in disease severity (Feng et al. 2012).

5.4.2.1   Targeting Expression of TNFAIP3 (A20)

Another regulator involved in the NF-kB activation pathway, known as TNF alpha-induced protein (TNFAIP3), also referred to as A20, participates actively in the innate immunity pathway. This protein codes for a ubiquitin-dependent enzyme and is expressed pleotropically in the cytoplasm. Interestingly, its expression is induced by NF-kB activation, and at the same time it also acts in a negative feedback loop and restricts its own expression. On the basis of GWAs studies, A20 has already been designated as a susceptibility gene for many autoimmune inflammatory disorders including IBD. The homoeostasis of A20 has been shown to get dysregulated in mucosal biopsies of CD patients at different stages of the disease where low expression is recorded. The expression of A20 regulates NF-kB signalling downstream of TNF receptor 1, Toll-like receptors (TLRs), NOD- like receptors and Interleukin-1 receptor (IL1-R). Recently, miR-125a, miR-125b, and miR-873 have been shown to activate the NF-kB pathway by targeting the expression of A20 (Figure 5.1).

5.4.2.2   Interfering in NOD2 Signaling

It has been demonstrated that upon stimulation with a bacterial product such as LPS, NOD2 interacts and initiates downstream signaling resulting in nuclear factor κB (NF-κB) activation and stimulation of downstream pro-inflammatory cytokines. NOD2 is designated as a susceptible gene for IBD. Functional studies using colonic epithelial cell lines, revealed miR-122 as one of the targets of NOD2, which activates NFkB pathway thereby, inhibiting the innate immune system (Chen et al. 2013). MiR-122 was also found to be deregulated with the progression of disease in CD patients. Further studies using colonic epithelial cell models have revealed four additional miRNAs that are able to attenuate innate immune response by suppressing NOD2 signaling. These were identified as miR-192, miR-495, miR-512 and miR-671 (Chuang et al. 2014). In MDP activated macrophages, miR-146a has been implicated to regulate the NOD2 downstream signaling resulting in gut inflammation in IBD and heightened expression of proinflammatory cytokines. It can be inferred from the above studies that NOD2 expression is regulated by a battery of micro RNAs. However, interestingly it is observed that the downstream immune responses and NOD2 expression did not correlate with each other and therefore, one can speculate

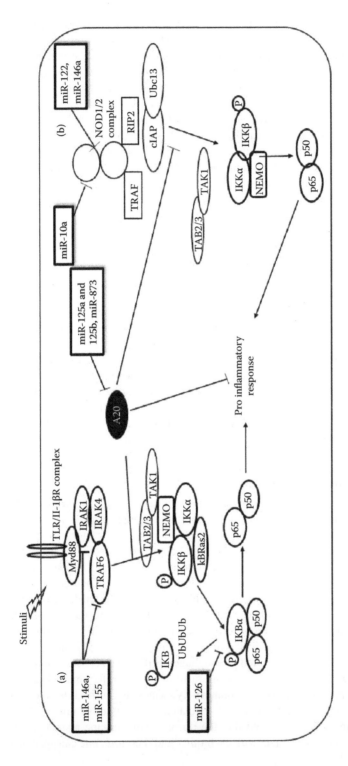

FIGURE 5.1 Representative figure showing the possible regulation of inflammatory pathway by microRNAs in IBD: (a) The TLR-mediated activation of NFkB results in the production of pro-inflammatory response. The important signaling molecules of this pathway, i.e. TRAF6 and IRAK1, are targeted by miR-146a. Also, the inhibitory subunit of NFkB, IKBa, is targeted by miR-126. (b) NOD2 receptor activation also results in downstream signaling, which is mediated by the proteosomal degradation of IkB, which is the inhibitory subunit of NFkB and NFkB activation. This pathway is also regulated by mIR-10a, which targets NOD2. A20 regulates both of these pathways by removing the K63-linked polyubiquitin chains from RIP2 and also negatively regulates the pro-inflammatory response. The expression of A20 is negatively regulated by miR-125a, 125b and miR 873.

that miRNAs might have multiple gene targets in downstream signaling pathways. Upon activation in dendritic cells, NOD2 is also capable of inducing expression of the family of miR-29 (29a, 29b and 29c) that in turn downregulate the expression of the target cytokine IL-23 (Brain et al. 2013). MiR-10a is also reported to get downregulated in the colonic mucosa of IBD patients, which is regulated by TNFα and IFNγ. Treatment with anti TNFα monoclonal antibodies resulted in increased miR-10a expression in CD patients. MiR-10a controls inflammation by targeting NOD2, IL-12 and IL-23p40 expression. An increased expression of these genes was observed in IBD patients, probably due to downregulation of miR-10a (Wu et al. 2015). High expression of miR-132 and miR-223 have also been reported in IBD patients, and when the regulatory mechanism was studied, both these miRNAs were found to target FOXO3 by bringing down its expression and thereby enhancing the NF-κB signaling which leads to overproduction of proinflammatory cytokines (Kim et al. 2016).

5.4.2.3   Interfering in T Cell Differentiation

MicroRNAs are known to affect the differentiation, maturation and activation of T and B lymphocytes which form the major arms of adaptive immunity. Specific deletion of Dicer and Drosha in T cell lineage causes unusual differentiation of T cells marked towards Th1 and excess production of IFNγ (Belver et al. 2011). MiR-146b, miR-29, miR-155, miR-128 and miR-27b are reported to effect the Th1 differentiation and function. MiR-155 overexpression facilitates the T cell differentiation towards Th1 cells, whereas its deficiency results in a shift towards Th2 subtype. Similar to this, miR-17-92 upregulates the production of IFNγ which induces the production of Th1 cells while suppressing Th2 subtype. Additionally, miR-21, which is well known for its upregulation during IBD, is shown to promote the production of Th2 lymphocytes. Some of the miRNAs modulate the expression of transcription factors that are responsible for Th1 expression where miR-29 targeting T-bet and eomesodermin known to induce IFNγ production and miR-146a targeting signal transducer and activator transcription 1 (STAT1) which is responsible for Treg mediated Th1 response. Besides this, regulatory T cells (Treg) and Th17 have been implicated to play important roles in the pathogenesis of autoimmune diseases including IBD. It has also been suggested that the imbalance in the population of Foxp3+CD4+ Treg and T effector cells (Th17) results in IBD pathogenesis (Yamada et al. 2016; Yao et al. 2012). Upon measuring Treg and Th17 cells in the peripheral blood of IBD patients, an increase in the population of Th17 while decrease in the Treg cells and the overall ratio of Treg/Th17 was also significantly decreased in the IBD patients (Eastaff et al. 2010). MicroRNAs are important for development and functioning of Treg cells as they promote the differentiation of CD4+ T cells in to Treg cells in thymus. *In vivo* studies have shown that CD4+ cells which do not express miRNAs develop spontaneous autoimmunity. MicroRNA such as miR-146a, miR-155, miR-17-92 and miR-10a modulate different signaling pathways and maintain Treg functions. *Yao* et al. reported that miR-155 regulates the Treg and Th17 cells differentiation by directly targeting SOCS1, the negative regulator of Treg and Th17 differentiation. It also enhances the production of proinflammatory cytokine IL-17A by Th17 cells (Yao et al. 2012). Additionally, miR-146a mediates the Treg mediated suppression of

Th1 response by targeting STAT1. Mir-146a deficient cells showed alteration in Treg suppressor function and impaired IFNγ response probably by increased expression and activation of STAT1 (Belver et al. 2011).

5.5 DYSREGULATION OF miRNA IN BIOPSY SAMPLES OF IBD PATIENTS

Previous studies had revealed differential expression of mRNA in different regions of the gastrointestinal tract. It has now been established that both the subtypes of IBD manifest an aberrant immune response to the gut flora of a genetically predisposed individual; however, the clinical presentation of the two diseases CD and UC differs. Between the two diseases, variations in the expression of inflammatory genes have been demonstrated. Differences have been recorded in the association studies of susceptibility genes conducted by GWAS. Set of miRNAs dysregulated in UC, and two subtypes of CD i.e. Crohn's colitis and Crohn's ileitis also did not overlap (Wu et al. 2011). The miRNA profile of different disease subtypes of IBD (i.e. ileal CD, colonic CD and UC) differs from each other which suggests that miRNA targets different inflammation related genes during these different disease stages. Studies focused on the changes in the expression pattern of miRNA with the progression of the disease; the aim being to speculate the role of miRNA as contributors of the disease. Identifying miRNAs with altered expression in IBD patients often found to co-localize with the genes which are identified as susceptibility genes for IBD. Studies have also shown that some miRNAs like miR-26a,-29a, -29b,-30c,-126*,-127-3p,-196a,-324-3p get dysregulated right at the quiescent state of the disease and altered levels continue during the active stage of IBD (Fasseu et al. 2010). Subsequently, it was shown that irrespective of the disease state, active or inactive, both UC and CD patients exhibit elevated level of miR-127-3p in the mucosal biopsy sample (Iborra et al. 2013) indicating potential use of miRNA as diagnostic biomarkers. Altered expression of miR-21 and miR-155 miR-20b and miR-7 has been reported in the colonic mucosa of patients with active UC (Takagi et al. 2010; Coskun et al. 2013; Nguyen et al. 2013). Two subtypes of disease can be distinguished based on the expression of identified miRs like miR-19a, miR-21, miR-31, miR-101, miR-146a, and miR-375 (Schaefer et al. 2015). Efforts to study the dysregulation of miRNA in different subtypes of patients, differing in disease behavior and localization of inflammation, have increased. Studies conducted on patients suffering from active UC in the sigmoid colon region demonstrated differential expression of miRNA compared to remission stages of the disease (Wu et al. 2010). A microarray profiling of dysregulated miRNA, conducted in north Indian patients of Ulcerative colitis revealed changes in the expression pattern based on the disease pathology and also varied with the site of the inflammation (Ranjha et al. 2015). Dysregulation of few miRNAs was quite pronounced in patients where the procto-sigmoid area was involved compared to patients where the ascending colon was involved during inflammation. However, the molecular basis of site-specific dysregulation of miRNA in different categories of UC and CD is yet to be demonstrated.

The site-specific studies indicate that the pattern of miRNA expression might be an indicator to assess the status of the disease and further detail study on the functional role of these miRNAs will pave the way for therapeutic interventions.

5.5.1 Differential Expression of miRNA in Inflamed Mucosa

MicroRNAs are differentially expressed in the inflamed colonic mucosa as compared to the non inflamed mucosa of IBD patients. Some of the recent studies have reported that miRNA behave differently in the endoscopically inflamed and non-inflamed regions of the colon. Altered expression of miR-21, miR-31, miR-101, miR-142-3p, miR-142-5p, miR-155, miR-223, miR-375, and miR-494 has been reported in the inflamed colonic mucosa of CD patients as compared to non inflamed mucosa. Similarly, miR-21, miR-101, miR-142-5p, miR-146a, miR-155, and miR-223 showed significant higher expression in the inflamed colonic mucosa form UC patients in comparison to non-inflamed one (Schaefer et al. 2015). In addition to this, miR-138, miR-708, miR-212, miR-223, miR-424, miR-148 are found to get upregulated in the inflamed colonic mucosa of UC patients with a fold change of more than ten folds. In a separate study conducted by Fasseau et al. (2010) in UC patients on biopsies from inflamed and non-inflamed regions revealed five miRNAs named miR-29a, miR-29b, miR-126*, miR-127-3p, miR324-3p in upregulated state whereas four miRNAs like miR-188-5p, miR-25, miR-320a, miR-346 in downregulated states both in patients suffering quiescent or active stage of the disease. Interestingly, significant dysregulation in the expression of miR-196a in quiescent stage onlys whereas four miRNAs (miR-7, miR-31, miR-135b, miR-223) only in active stage of UC. However, in the case of CD patients, differential expression of miRNA was observed based on the site of the disease. In the case of Crohn's colitis patients, expression of miR-23b, miR-106a and miR-191 were at elevated levels, whereas decreased expression was observed for miR-19b and miR-609. In case of Crohn's ileitis patients, miR-16, miR-21, miR-223, and miR-594 levels were found to be elevated. Analysis of colon biopsy samples from UC patients of Norther India, using a microarray platform, revealed differential expression of miRNA based on different affected sites and different stages of the disease (Figure 5.2).

5.5.2 miRNA Associated with Epithelial Barrier Function Contributes in the Pathogenesis of IBD

To understand the pathophysiology of IBD comprehensively, the study was carried out with large IECs from a mouse colitis model and gave an insight into the barrier function and complexity of miRNA networks that arise with the severity of the disease. During severe conditions, in the inflamed large -IECs, number of miRNAs get dysregulated like miR-1224-5p, targeting ATP binding cassette transporter (ABCG2), miR-3473a targeting aquaporin 8 (AQP8). These targets are known to be involved in the progression of the disease (Lee et al. 2015).

Genome-wide association studies have revealed that two important IBD associated genes responsible for maintaining the epithelial barrier integrity are LAMB-1,

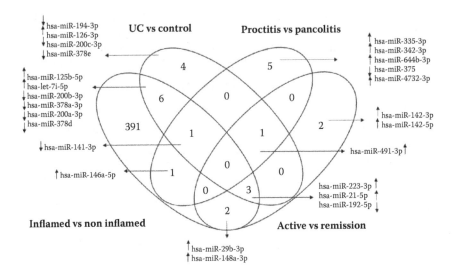

FIGURE 5.2 Venn diagram showing the microRNA dysregulated during different stages of ulcerative colitis in North Indian patients.

regulating the membrane stability, and CDH-1, maintaining the tight junctions with the help of E-cadherin. MiR-192, miR-21, miR-200b, miR-155, miR-126, miR-106a are identified to be involved in the maintenance and integrity of the epithelial barrier. Studies carried out by Wu et al. (2008) confirmed that there was a 47.1% decrease in the expression of miR-192 in the inflamed colon tissue of active UC patients compared to controls. Macrophage inflammatory peptide (MIP)-2 alpha, a CXC chemokine, also expressed by epithelial cells, is targeted by miR-192. Mir-21 is known to play a key role in the pathogenesis of both UC and ileal CD patients and demonstrated to have epithelial barrier function. Using dextran sodium sulfate induced colitis model of mice, it was shown that miR-21 knockout mice exhibited reduced level of apoptosis and intestinal permeability compared to wild type (Shi et al. 2013). Similarly, upregulated expression of miR-200b contributes in the pathogenesis of IBD as it targets the transforming growth factor β1 and inhibits epithelial-mesenchymal transition leading to loss of epithelial cells and dislodges the epithelial barrier integrity. During the process of inflammation of intestinal mucosa, intestinal stromal cells like intestinal fibroblasts and myofibroblasts (IMFs) get modulated and deregulate cytokine release. In case of UC, miR-155 modulates the inflammatory phenotypes of IMFs by targeting suppressor of cytokine signaling 1 (SOCS1) (Pathak et al. 2015). It is well known that both intracellular cell adhesion molecule-1 (ICAM-1) and vascular cell adhesion molecule-1 (VCAM-1) are expressed on the surface of fibroblasts. High expression of adhesion molecules in the tissue help in the initiation of leukocyte migration and local inflammation. Expression of VCAM-1 has been reported to be quite pronounced in the patient's samples of both UC and CD. Mir-126 has been reported to target VCAM-1 (Harris et al. 2008).

5.6   miRNA MEDIATING INTERACTION BETWEEN MICROBES AND HOST MUCOSAL IMMUNE RESPONSE DURING IBD

Role of gut microbiota in the pathogenesis of IBD has been established. MiR-10a, besides regulating the chronic intestinal inflammation, is known to maintain intestinal homoeostasis. Expression of miR-10a is negatively regulated by gut bacteria in the dendritic cell, and thus, the expression of cytokines IL-12/IL-23 is affected in these cells. It is suggested that commensal bacteria also regulated the expression of IL-23 in macrophages by inhibiting the expression of miR-107. This is carried out through the interaction of TLR-TLR ligands. Expression of PepT1, a di/tripeptide transporter belonging to a member of solute transporter family (SLC15A1), has been found in an elevated level in the colonic tissue of UC patients. PepT1 is responsible for uptake of bacterial products like muramyl-dipeptides into the cytosol of colonic tissue. This brings about a paradigm shift from symbiotic relationship to pathogenic relationship between the gut microbiota and the host during progression of the disease. Experiments carried out to see the differential expression of miRNA during active UC revealed that miR-193a was highly deregulated as compared to control. This inverse relationship between PepT1 and miR-193a was further confirmed by in vivo studies using DSS induced mice model of colitis. Restoration of miR-193a improved the inflammation (Xin et al. 2015).

5.7   miRNA IN IBD ASSOCIATED COLORECTAL CANCER (CRC)

Nearly 5.6% of patients suffering from Crohn's colitis and nearly 30% of patients suffering from UC are likely to develop colorectal cancer if the disease continues for 6–8 years of duration. Transcriptional regulation during the sporadic CRC and IBD, associated CRC are expected to be different. It is important to carry out proper surveillance throughout the process of transformation from IBD-dysplasia-CRC. Often monitoring different stages of the disease by histopathological examinations are difficult, therefore, molecular tools are developed for the purpose. Transcriptional alterations in the mucosal surface during the progression of neoplasia to invasive cancer are now being explored for early diagnosis (Colliver et al. 2006). Besides the role of miRNAs in inflammation during IBD, they have been reported to play a role in neoplastic tissue progression in inflammatory bowel disease associated colorectal cancer. Kanaan et al. reported down regulation of five miRNAs in the progression of non neoplastic tissues to dysplasia and from dysplasia to cancer. In CD patients, six miRNAs exhibited differential expression in non-neoplastic tissues, dysplasia and cancer. MiR-181a, miR-146b-5p, let-7e, miR-17 were upregulated during non-neoplastic to dysplasia; however, let-7e, miR-17 and miR-143 showed down regulation from dysplasia to cancer progression (Kanaan et al. 2012). A possible salvage mechanism has been proposed to explain the net down regulation of the above miRNAs during transformation from dysplasia to cancer where reversal of tumorigenesis is attempted by the affected cells.

5.8   DYSREGULATION OF CIRCULATING miRNA IN IBD PATIENTS

The main purpose of evaluating the circulating miRNAs during disease condition has been to establish a semi-invasive method of assessing the status of the disease conditions by measuring the differential expression of miRNAs that have a potential to be used as biomarkers. It has now been demonstrated that miRNAs are observed in extracellular spaces; they may be present in different biological fluids such as tears, breast milk, urine, saliva, serum etc. In these fluids they are found to be packaged in vesicles containing microparticles, lipoproteins and RNA binding proteins, thus they are protected from degradation (Chen et al. 2008). Wu et al. (2011) observed for the first time that the circulating immune cells detected during disease conditions in UC and CD patients are regulated to a great extent by differential expression of selected miRNAs when compared with control individuals. Among the subtypes of IBD, in case of UC selective increased expression of ten miRNAs and decreased expression of one miRNA were observed compared to CD patients. This study indicated that dysregulation of miRNAs varies on the subtype of IBD and can be explored to diagnose the disease subtypes in IBD (Wu et al. 2011). In the active stage of the IBD, one can observe increased expression of inflammatory genes due to downregulation in expression of many miRNA species. The profiling of circulating miRNAs in IBD patients would provide valuable information regarding the pathophysiology of the disease. Research is on to evaluate specific miRNAs from serum or peripheral blood that are differentially expressed in UC or CD at different stages of the disease. Peripheral blood samples are commonly used as genetic screens for various diseases and to identify various miRNA markers for diagnosing different cancers. The idea was if one can replicate the same miRNAs that are differentially regulated in biopsy samples. Zham et al. (2011) demonstrated altered miRNA expression in rectal and serum samples of pediatric IBD patients. They reported up regulation of eleven miRNAs in the pediatric CD patients and a significant decrease was observed in the expression of these miRNAs when they were maintained on remission for six months (Zahm et al. 2011). Another study identified miR-155 as a most highly expressed miRNA in blood samples of UC patients, whereas in the case of CD, higher expressions were noticed for miR-16, miR-23a, miR-29a, miR-106a, miR-107, miR-126, miR-191, miR-199a-5p, miR-200c, miR-362-3p and miR-532-3p (Paraskevi et al. 2012). Schaefer et al. reported a number of miRNAs (miR-19a, miR-21, miR-31, miR-101, miR-146a, and miR-375) that can be used as discriminatory markers for identifying CD and UC. Most of these miRNA expressions were replicated in extra intestinal samples such as peripheral blood and saliva samples (Schaefer et al. 2015). A study was conducted to predict the incidences of UC by Duttagupta et al. on genome maps of circulating miRNAs that are present in micro-vesicles, platelets and PBMCs. They identified 31 differentially expressed microRNAs in the blood platelets and subjected them to rigorous tests to assess their biomarker potentials. These signature miRNAs were further validated to 88% accuracy through qPCR assays. Target prediction tools provided better insights into the biological relevance of this miRNA dysregulation by predicting their putative target genes. (Duttagupta et al. 2012). Table 5.1 enlists various miRNA reported to be dysregulated when samples were analysed either from biopsy or serum samples of IBD patients.

TABLE 5.1
List of MicroRNA Dysregulated during UC and CD

MicroRNA	Sample	Approach	Disease	Reference
miR-192,375,422b,16, 21,23a,24,29a,126,195.let-7f	Colon biopsy	Microarray qRT-PCR	Active UC	Wu et al. 2008
miR-21, 155	Colon biopsy	Microarray qRT-PCR	Active UC	Takagi et al. 2010
miR-19b.629,23b,106a,191 16,21,223,594	Colon biopsy	Microarray qRT-PCR	Active CD	Wu et al. 2008
miR188,215,320a,346,7,31,135b 223,29a,29b,126,127,324	Colon biopsy	qRT-PCR	Active UC	Fasseu et al. 2010
miR9,126,130a,181c,375,26a 29b,30b,34c,126,127,133b,155, 196a, 324,21,22,29c,31,106a,146a, 146b,150	Colon biopsy	qRT-PCR	Active CD	Fasseu et al. 2010
miR9,30a,30c,223,26a,29b,30b, 34c,126,127,133,155,196a,324,21 22,29c,31.106a.146a,146b,150	Colon biopsy	qRT-PCR	Inactive CD	Fasseu et al. 2010
miR150,196b,199a,199b, 223,320a	Colon biopsy	qRT-PCR	Inactive UC	Fasseu et al. 2010
miR-196	Colon biopsy	qRT-PCR, ISH	Active CD	Brest et al. 2011
miR23a,106a,107,126,191 199a,362,532	Blood	qRT-PCR	Active CD	Paraskevi et al. 2012
miR-199a,28,151	Blood	qRT-PCR	Active UC	Paraskevi et al. 2012
miR-27a,140,195,877	Blood	qRT-PCR	Active CD	Iborra et al. 2013
miR-760,423,103,15b,93,598 374b,30e,28	Blood	qRT-PCR	Active UC	Iborra et al. 2013
miR-20b	Colon biopsy	Microarray qRT-PCR	Active UC	Coskun et al. 2013
miR-7	Colon biopsy	qRT-PCR	Active CD	Nguyen et al. 2010

5.9   SNP DISCOVERY IN miRNA FROM IBD PATIENTS

Single nucleotide polymorphism (SNP) in miRNA influences its expression and maturation. If the SNP is present within the seed region, it hampers the binding of miRNA with its target gene and results in an altered miRNA:mRNA interaction. Similarily, SNP in the 3'UTR of a gene could induce or delete the binding site for miRNA, which eventually results in the altered expression of the respective gene and affects disease pathogenesis. The functional variants of miRNA have been identified in cases of different human diseases, including IBD, and have been found to contribute in disease pathogenesis. SNP studies conducted in different ethnic populations showed variation in disease susceptibility. Okubo et al. demonstrated the association of three SNPs (rs1614913, rs2910164, rs3746444) in pre-miRNAs coding regions (miR-196a2, miR-146a and miR-499) with disease susceptibility in

IBD in the Japanese population (Okubo et al. 2011). When demographic features were correlated, it was observed that AG genotype of rs3746444 in miR-499 correlated well with onset of disease at an older age, left-sided colitis, pan colitis, steroid dependence, frequency of hospitalization and refractoriness towards medical therapy; whereas, patients with AA genotype of the same miRNA were inversely correlated with the above demographic features. However, patients with TT genotype of miRNA-196a2 (rs1614913) were at higher risk since these patients were completely refractory to therapy.

Rs3746444 AG genotype was observed to be significantly higher in UC patients (OR = 1.51, p = 0.037), however, there were no significant changes between UC and controls in the case of SNP rs11614913 and rs2910164. Gazauli et al. reported the association of two SNPs (rs2910164 and rs11614913) in miR-146a and miR-196a2 with IBD susceptibility in Greek population. In the case of miR-146a rs2910164, the CC genotype and C allelic frequencies were found to be significantly higher in CD patients as compared to controls (p = 0.001, OR-4.51 and p = 0.0001, OR = 2.21 respectively), but this association was not observed in UC. While in case of miR-196a2 rs11614913, a protective role of TT genotype and T allele towards UC was suggested (p = 0.019, OR = 0.48 and p = 0.008, OR = 0.69, respectively), however, these two SNPs did not show any phenotypic change in UC and CD patients of Greek population (Ciccacci et al. 2016). MiR-196a-2 rs11614913 (C>T) and miR-499 rs3746444 (T>C) SNPs were analyzed in UC patients of North Indian populations. Interestingly the homozygous mutant of MiR-196a (TT genotype) was found to be negatively associated whereas the heterozygous mutant of miR-499 (TC genotype) exhibited a positive association. When demographic features were compared, patients harboring both the mutations in the above miRNAs, developed disease at an older age with high severity. Further, it was observed that the genotypes showing association with the disease also correlated with the changes in expression of their respective miRNAs (Ranjha et al. 2017).

Brest et al. (2011) reported that a synonymous variant of Immunity related GTPase M (IRGM) gene which is a susceptibility gene for CD was in a strong linkage disequilibrium with a deletion polymorphism. Based on the hypothesis that polymorphism could hamper the miRNA mediated suppression of target mRNA, the binding of miR-196 with IRGM was investigated, and it was found that miR-196 significantly altered the expression of the IRGM protective allele (c.313C) but did not target the risk associated allele (c.313T) (Brest et al. 2011). IL-23 is well known to be a key player in IBD pathogenesis as it plays critical role in the Th1-IFN-γ inflamatory axis in the intestine and is also known to inhibit the deveolment of FoxP3+ Treg cells which have an important role in controling the inflammatory response. A functional variant of IL-23R has been reported as a risk factor for both subtypes of IBD, ulcerative colitis and Crohn's Disease. SNP rs10889677 in the 3'UTR of this gene leads to an increase in expression of this gene both at mRNA and protein levels. When co-expressed with miR-let7e and let-7f, these microRNA were able to suppress the expression of wild type gene but not the variant one, implicating that the SNP in the 3' UTR resulted in loss of miRNA binding site which suggested that along with other genetic risk factors this mutation could act as a contributing factor in pathogenesis by sustaining the IL-23 signaling (Zwiers et al. 2012).

5.10   DIAGNOSTIC POTENTIALS OF miRNA
AND ROLE AS TARGETS IN THERAPEUTICS

Available research data clearly indicates involvement of miRNA in the pathogenesis of IBD. Therefore, miRNAs are explored for their application as therapeutics. From various studies, it is now evident that miRNAs can be isolated from various samples beside mucosal tissues, like blood or serum samples, saliva of the body, etc. Many of these procedures are much less invasive, making them suitable for diagnosis. These miRNAs are also found to be biologically active and stable and therefore can be exploited for diagnostic and therapeutic purposes.

It is now understood that decreased expression of miR-192 is experienced during active CD; therefore, its therapeutic role has been investigated. Upon intravenous administration of an antagomir in a mice model to silence the miR-192 there was a reduction in the level of the micro RNA in intestinal tissues. This clearly indicates the possibility of developing a tissue-specific delivery system that can prevent unintended off-target effects (Krutzfeldt et al. 2005).

Disease activity index can also be assessed by profiling miRNA at different stages of the disease. Traditionally, C-reactive protein value is considered as a biomarker for assessing the disease activity, however, recently a panel of circulating miRNAs has been developed (miR-4454, miR-223-3p, miR-23a-3p) that correlates well with UC activity and exhibits better sensitivity than C-reactive protein value. Streamlining the techniques for isolation of miRNAs from the site of inflammation, or from the peripheral tissues, will generate data that will determine the efficacy of the miRNAs to be used as therapeutics. Once the specific targets of miRNA are identified, and their functional roles are determined, development of miRNA based therapy is proposed for in vivo applications. This development consists of designing the mimics or even inhibitors based on synthetically designed oligonucleotides and their delivery system. The basis of designing anti-microRNA oligonucleotides (AMOs) consists of the following steps (Chapman and Pekow 2015):

1. Constructing synthetic oligonucleotides based on the complementary sequence of the mature miRNA.
2. The sequence should be highly specific, stable and must bind to the target miRNA.
3. It should meet the pharmacokinetics required to block the function of the targeted miRNA.
4. It needs to be chemically modified to prevent nuclease activity when delivered in the blood stream.
5. It needs to be optimized for smooth cellular uptake.

The modifications made so far are on the phosphate backbone or sugar moiety or introducing nonribose backbones. The sugar modifications tried so far include antisense 2'-O-methyl (2'-OMe) and 2'-O-methoxyethyl (2'-MOE) oligoribonucleotides, and 2', 4'-methylene bridge. These lock nucleic acid (LNA) based molecules improve the binding affinity and are able to reduce degradation (Lennox and Behlke 2011).

Further modifications have been made by substituting a sulfur atom to the phosphate backbone. This helps in prevention of nuclease mediated degradation.

In order to design targets against miRNAs grouped in a family (as they share common seed sequences), antisense AMOs, which are synthetically designed, offer limited target depth, and therefore, miRNA sponges are designed. These sponges basically consist of plasmid constructs with a strong promoter. These constructs contain multiple tandem binding sites for target mRNA for miRNAs of interest. Therefore, it facilitates competitive inhibition of multiple miRNAs (Lu et al. 2009). However, *in vivo* applications require further validation to avoid any untoward side effects. Mir masks have been designed to inhibit the action of miRNA where antisense oligonucleotides (ASO) are developed, which is a modified version of AMO. In this case, single stranded 2'-OMe modified oligoribonucleotide is designed so as to bind to the 3'-UTR of the target mRNA by its complementary sequence, and thus blocks the access of miRNAs to the target mRNA, thus able to repress the activity in a gene specific manner. These have been applied only *in vitro* models so far. MiR-155 is known to get dysregulated during UC and in other inflammatory disorders. MiR-155 specific AMOs were designed using a locked nucleic acid (LNA), antimiR-155 repressed the granulocyte-colony stimulating factor in activated macrophages during acute inflammatory responses both *in vitro* and *in vivo*. (Worm et al. 2009).

Among other ways of miRNA therapeutics, *miRNA mimics* are constructed as gene silencing approaches. Double-stranded RNA are artificially generated where one strand contains a sequence motif identical to the endogenous miRNA and pair on its 5'-end that is partially complementary to the target sequence in the 3'UTR, the passenger strand, required for loading into the RISC complex. In order to prevent nuclease digestion and to facilitate cellular uptake, chemical modifications such as 2'-OMe or cholesterol conjugation have been proposed. Other than this, *miRNA expression gene vectors* have been developed where DNA plasmids or viral vectors consisting of adeno-associated viral vectors (AAV) are constructed that can be used to express miRNA constructs constitutively. These vectors are most suitable as they are nonpathogenic in nature and can offer high level expression and transduction efficiency. AAV vectors have proven to be useful in experimental model of IBD by introducing the vector containing IL-10 in TNBS induced colitis in mice or IL-10 knockout mice. (Lindsay et al. 2003). In both the cases amelioration of inflammation were observed. Therefore, one can speculate that use of miRNA mimics containing similar sequence identity with the endogenous miRNAs may provide an alternative therapeutic approach to reestablish the pathways that are dysregulated during IBD. In an epithelial cell culture model, TNFα induced MIP-2a expression was successfully inhibited by miR-192 mimic (Wu et al. 2008).

It is well known that miR-21 gets highly upregulated in the colonic tissue of UC patients. In an *ex vivo* colon epithelial cell model, miR-21 mimics caused loss of tight junction proteins and also resulted in an increased barrier permeability (Yang et al. 2013). ATG16L1 responsible for autophagic activity has been identified as one of the susceptible gene in case of CD. Using colon cell lines, Zhai et al. (2013) and Lu et al. (2014) established that miR-106b and miR-93 regulate the expression of ATG16L1 where it binds to the 3'-UTR of the gene. This was further confirmed in colonic tissue of CD patients where increased expression of miR-106b was observed with simultaneous reduction in ATG16L1 expression. To perform this investigation, they used anti-mRNA oligonucleotides (AMOs) against miR-106b and miR-93 as well as mimics of miR-106b and miR-93. All the above mentioned techniques can be applied provided the dysregulated miRNAs are identified in a specific disease condition.

5.11 CONCLUSION

The role of micro RNA as epigenetic regulator in the pathogenesis of IBD has been established by painstaking research work over the last decade. The advent of advanced computational, biochemical and genetic techniques for scoring the expression of micro RNAs and their targets expanded our knowledge to understand their mechanism of action at different stages of the disease. Since they add a layer of complexity in the regulation of genes involved in the inflammatory process, investigation focused on possible use of miRNAs as a biomarker for diagnostics and therapeutics. Experiments carried out to study miRNA expression profiling both from biopsy samples and peripheral blood of UC and CD patients has unraveled the functional pathways where miRNA exerts post transcriptional regulation of target genes. The studies so far carried out face issues like variations in the miRNA expressions, often due to small sample sizes and heterogeneity that exists within the patients and patient subtypes. Therefore, more research is needed to standardize the protocol for quantification of miRNAs isolated from different sources like peripheral blood or tissue biopsies. Current research has also focused on the role of miRNAs in the epithelial barrier function by targeting vital genes responsible for maintaining the integrity of IECs, tissue homoeostasis and cell differentiation. Functional studies on miRNA using *in vitro* and *in vivo* models have shown overlapping with the susceptible genetic loci of IBD detected earlier by GWAS studies. Single nucleotide polymorphisms have been detected in some miRNAs in the seed region which may disturb the miRNA:mRNA interaction, thus interfering its expression on the target genes involved in the inflammatory pathways. In some patients, during the progression of UC or CD towards colorectal cancer, deregulated expression of miRNA has been identified responsible for the transition. Overall the available literature indicates the potential of miRNA for clinical management of disease and also adds knowledge to develop therapeutics by designing suitable antimirs.

REFERENCES

Adams AT, Kennedy NA, Hansen R et al. 2014. Two-stage genome-wide methylation profiling in childhood-onset Crohn's disease implicates epigenetic alterations at the VMP1/MIR21 and HLA loci. *Inflamm Bowel Dis* **20**:1784–1793.

Ananthakrishnan AN. Epidemiology and risk factors for IBD. 2015. *Nat Rev Gastroenterol Hepatol* **12**(4): 205–217.

Belver L, Papavasiliou FN, Ramiro AR. 2011. MicroRNA control of lymphocyte differentiation and function. *Cur Opin Immunol* **23**(3): 368–373.

Brain O, Owens B, Pichulik T et al. 2013. The intracellular sensor NOD2 induces microRNA-29 expression in human dendritic cells to limit IL-23 release. *Immunity* **39**: 521–536.

Brest P, Lapaquette P, Souidi M et al. 2011. A synonymous variant in IRGM alters a binding site for miR-196 and causes deregulation of IRGM-dependent xenophagy in Crohn's disease. *Nat Genet* **43**(3): 242–245.

Chapman CG, Pekow J. 2015. The emerging role of miRNAs in inflammatory bowel disease: A review. *Ther Adv Gastroenterol* **8**(1): 4–22.

Chen X, Ba Y, Ma L, Zhang Y et al. 2008. Characterization of microRNAs in serum: A novel class of biomarkers for diagnosis of cancer and other diseases. *Cell Res* **18**(10): 997–1006.

Chen Y, Wang C, Liu Y et al. 2013. MiR-122 targets NOD2 to decrease intestinal epithelial cell injury in Crohn's disease. *Biochem Biophys Res Comm* **438**: 133–139.

Chen Y, Xiao Y, Ge W et al. 2013. MiR-200b inhibits TGF-β1-induced epithelial-mesenchymal transition and promotes growth of intestinal epithelial cells. *Cell Death Dis* **4**: e541.

Cheng X, Zhang X, Su J et al. 2015. miR-19b downregulates intestinal SOCS3 to reduce intestinal inflammation in Crohn's disease. *Sci Rep* **5**: 10397, doi: 10.1038/srep10397

Cho JH. 2008. The genetics and immunopathogenesis of inflammatory bowel disease. *Nat Rev Immunol* **8**(6): 458–466.

Chuang A, Chuang J, Zhai, Z et al. 2014. NOD2 expression is regulated by microRNAs in colonic epithelial HCT116 cells. *Inflamm Bowel Dis* **20**: 126–135.

Ciccacci C, Politi C, Novelli G et al. 2016. Advances in exploring the role of microRNAs in inflammatory bowel disease. *MicroRNA* **5**(1): 5–11.

Colliver DW, Crawford NP, Eichenberger MR et al. 2006. Molecular profiling of ulcerative colitis-associated neoplastic progression. *Exp Mol Pathol.* **80**: 1–10.

Coskun M, Bjerrum JT, Seidelin JB et al. 2013. miR-20b, miR-98, miR-125b-1*, and let-7e* as new potential diagnostic biomarkers in ulcerative colitis. *World J Gastroenterol* **19**(27): 4289–4299.

Darfeuille-Michaud A, Boudeau J, Bulois P et al. 2004. High prevalence of adherent-invasive *Escherichia coli* associated with ileal mucosa in Crohn's disease. *Gastroenterology* **127**(2): 412–421.

Duttagupta R, DiRienzo S, Jiang R et al. 2012. Genome-wide maps of circulating miRNA biomarkers for ulcerative colitis. *PloS One* **7**(2): e31241.

Eastaff-Leung N, Mabarrack N, Barbour A et al. 2010. Foxp3+ regulatory T cells, Th17 effector cells, and cytokine environment in inflammatory bowel disease. *J Clin Immunol* **30**(1): 80–89.

Fasseu M, Treton X, Guichard C et al. 2010. Identification of restricted subsets of mature microRNA abnormally expressed in inactive colonic mucosa of patients with inflammatory bowel disease. *PloS One* **5**(10): e13160.

Feng X, Wang H, Ye S et al. 2012. Up-regulation of microRNA-126 may contribute to pathogenesis of ulcerative colitis via regulating NF-kappaB inhibitor IkappaBalpha. *PloS One* **7**(12): e52782.

Geremia A, Biancheri P, Allan P et al. 2014. Innate and adaptive immunity in inflammatory bowel disease. *Autoimmun Rev* **13**(1): 3–10.

Harris TA, Yamakuchi M, Ferlito M et al. 2008. MicroRNA-126 regulates endothelial expression of vascular cell adhesion molecule 1. *Proc Natl Acad Sci* **105**(5): 1516–1521.

Iborra M, Bernuzzi F, Correale C et al. 2013. Identification of serum and tissue microRNA expression profiles in different stages of inflammatory bowel disease. *Clin Exp Immunol* **173**(2): 250–258.

Kanaan Z, Rai SN, Eichenberger MR et al. 2012. Differential microRNA expression tracks neoplastic progression in inflammatory bowel disease-associated colorectal cancer. *Hum Mut* **33**(3): 551–560.

Kim HY, Kwon HY, Ha Thi HT et al. 2016. MicroRNA-132 and microRNA-223 control positive feedback circuit by regulating FOXO3a in inflammatory bowel disease. *J Gastroenterol Hepatol* **31**(10): 1727–1735.

Kozomara A, Griffiths-Jones S. 2011. miRBase: Integrating microRNA annotation and deep-sequencing data. *Nucleic Acids Res* **39**(Database issue): D152–D157.

Krutzfeldt J, Rajewsky N, Braich R et al. 2005. Silencing of microRNAs *in vivo* with 'antagomirs'. *Nature* **438**: 685–689.

Lee J, Park EJ, Yuki Y et al. 2015. Profiles of microRNA networks in intestinal epithelial cells in a mouse model of colitis. *Sci Rep* **5**:18174; doi: 10.1038/srep18174

Lennox KA, Behlke MA. 2011. Chemical modification and design of anti-miRNA oligo-nucleotides. *Gene Ther* **18**(12): 1111–1120.

Lindsay JO, Ciesielski CJ, Scheinin T et al. 2003. Local delivery of adenoviral vectors encoding murine interleukin 10 induces colonic interleukin 10 production and is therapeutic for murine colitis. *Gut* **52**(7): 981–987.

Loddo I, Romano C. 2015. Inflammatory bowel disease: Genetics, epigenetics, and pathogenesis. *Front Immunol* **6**: 551.

Lu C, Chen J, Xu H et al. 2014. miR106b and miR93 prevent removal of bacteria from epithelial cells by disrupting ATG16l1-mediated autophagy. *Gastroenterology* **146**: 188–199.

Lu Y, Xiao J, Lin H et al. 2009. A single anti-microRNA antisense oligodeoxyribonucleotide (AMO) targeting multiple microRNAs offers an improved approach for microRNA interference. *Nucleic Acids Res* **37**(3): e24.

Nguyen HT, Dalmasso G, Yan Y et al. 2010. MicroRNA-7 modulates CD98 expression during intestinal epithelial cell differentiation. *J Biol Chem* **285**(2): 1479–1489.

Okubo M, Tahara T, Shibata T et al. 2011. Association study of common genetic variants in pre-microRNAs in patients with ulcerative colitis. *J Clin Immunol* **31**(1): 69–73.

Paraskevi A, Theodoropoulos G, Papaconstantinou I et al. 2012. Circulating microRNA in inflammatory bowel disease. *J Crohns Colitis* **6**(9): 900–904.

Pathak S, Grillo AR, Scarpa M et al. 2015. MiR-155 modulates the inflammatory phenotype of intestinal myofibroblasts by targeting SOCS1 in ulcerative colitis. *Exp Mol Med* **47**: e164.

Ranjha R, Aggarwal S, Bopanna S et al. 2015. Site-specific microRNA expression may lead to different subtypes in ulcerative colitis. *PloS One*, **10**(11): e0142869.

Ranjha R, Meena NK, Singh A et al. 2017. Association of miR-196a-2 and miR-499 variants with ulcerative colitis and their correlation with expression of respective miRNAs. *PloS One* **12**(3): e0173447

Scarpa M, Stylianou E. 2012. Epigenetics: Concepts and relevance to IBD pathogenesis. *Inflam Bowel Dis* **18**(10): 1982–1996.

Schaefer JS, Attumi T, Opekun AR et al. 2015. MicroRNA signatures differentiate Crohn's disease from ulcerative colitis. *BMC Immunol* **16**: 5.

Shi C, Liang Y, Yang J et al. 2013. MicroRNA-21 knockout improve the survival rate in DSS induced fatal colitis through protecting against inflammation and tissue injury. *PLoS One* **8**: e66814.

Szucs D, Beres NJ, Rokonay R et al. 2016. Increased duodenal expression of miR-146a and -155 in pediatric Crohn's disease. *World J Gastroenterol* **22**(26): 6027–6035.

Taganov KD, Boldin MP, Chang KJ et al. 2006. NF-kappaB-dependent induction of microRNA miR-146, an inhibitor targeted to signaling proteins of innate immune responses. *Proc. Natl Acad Sci* **103**(33): 12481–12486.

Takagi T, Naito Y, Mizushima K et al. 2010. Increased expression of microRNA in the inflamed colonic mucosa of patients with active ulcerative colitis. *J Gastroenterol Hepatol* **25** Suppl 1: S129–S133.

Worm J, Stenvang J, Petri A et al. 2009. Silencing of microRNA-155 in mice during acute inflammatory response leads to derepression of C/EBP beta and down-regulation of G-CSF. *Nucleic Acids Res* **37**: 5784–5792.

Wu F, Guo NJ, Tian H et al. 2011. Peripheral blood microRNAs distinguish active ulcerative colitis and Crohn's disease. *Inflamm Bowel Dis* **17**(1): 241–250.

Wu F, Zhang S, Dassopoulos T et al. 2010. Identification of microRNAs associated with ileal and colonic Crohn's disease. *Inflamm Bowel Dis* **16**(10): 1729–1738.

Wu F, Zikusoka M, Trindade A et al. 2008. MicroRNAs are differentially expressed in ulcerative colitis and alter expression of macrophage inflammatory peptide-2 alpha. *Gastroenterology* **135**(5): 1624–1635.

Wu W, He C, Liu C et al. 2015. miR-10a inhibits dendritic cell activation and Th1/Th17 cell immune responses in IBD. *Gut* **64**(11): 1755–1764.

Xavier RJ, Podolsky DK. 2007. Unravelling the pathogenesis of inflammatory bowel disease. *Nature* **448**: 427–434.

Yamada A, Arakaki R, Saito M et al. 2016. Role of regulatory T cell in the pathogenesis of inflammatory bowel disease. *World J Gastroenterol* **22**(7): 2195–2205.

Yang L, Boldin MP, Yu Y et al. 2012. miR-146a controls the resolution of T cell responses in mice. *J Exp Med* **209**(9): 1655–1670.

Yang Y, Ma Y, Shi C et al. 2013. Overexpression of miR-21 in patients with ulcerative colitis impairs intestinal epithelial barrier function through targeting the Rho GTPase RhoB. *Biochem Biophys Res Commun* **434**: 746–752.

Yao R, Ma YL, Liang W et al. 2012. MicroRNA-155 modulates Treg and Th17 cells differentiation and Th17 cell function by targeting SOCS1. *PloS One*, **7**(10): e46082.

Zahm AM, Thayu M, Hand NJ et al. 2011. Circulating microRNA is a biomarker of pediatric Crohn disease. *J Ped Gastroenterol Nutr* **53**(1): 26–33.

Zhai Z, Wu F, Chuang A et al. 2013. miR-106b fine tunes ATG16l1 expression and autophagic activity in intestinal epithelial HCT116 cells. *Inflamm Bowel Dis* **19**: 2295–2301.

Zwiers A, Kraal L, van de Pouw Kraan TC et al. 2012. Cutting edge: A variant of the IL-23R gene associated with inflammatory bowel disease induces loss of microRNA regulation and enhanced protein production. *J Immunol* **188**(4): 1573–1577.

6 Involvement of MicroRNAs in Alzheimer's Disease

Malay Bhattacharyya and
Sanghamitra Bandyopadhyay

CONTENTS

Introduction

The microRNAs (miRNAs) are a class of endogenous, short, non-coding RNAs of length around 22nt, which are involved in post-transcriptional regulation of mRNAs (Bartel, 2004, 2009). The first miRNA discovered in 1993 by Ambros et al. in *Caenorhabditis elegans* was of the let-7 family. Ever since then miRNAs have also been documented in various other eukaryotes. They are presently one of the most widely studied non-coding regulatory molecules across eukaryotes. Recent studies reveal the dual annotation of miRNAs with other non-coding RNAs, thereby increasing their importance. Over the last two decades, there has been an exponential growth of literature establishing the involvement of miRNAs in different kinds of diseases (Chen, 2005).

The regulation of and by miRNAs is still a growing field of research. Regulation by miRNAs is explicable to some extent, but their own regulation is not entirely clear. This is mainly because miRNAs can be categorized into two types. Depending on their genomic loci, they are classified as intragenic miRNAs and intergenic miRNAs (Bhattacharyya et al., 2012). The former is transcribed by RNA Pol II promoter of its host genes whereas the latter is transcribed by independent RNA Pol II promoters (Corcoran et al., 2009). The biogenesis of miRNAs includes the successive generation of primary miRNAs, precursor miRNAs and mature miRNAs (Kim, 2005). The mature miRNAs combine with AGO subfamily of argonaute proteins to form the RNA Induced Silencing Complex to post-transcriptionally inhibit gene expression. However interestingly, some recent studies also reveal the novel involvement of primary miRNAs in the functions like repression of genes (Trujillo et al., 2010) and peptide encoding (Lauressergues et al., 2015). There exists a complex regulatory network between the transcription factors, miRNAs and genes (Bandyopadhyay and Bhattacharyya, 2009).

Alzheimer's disease (AD) is a kind of neurodegenerative disease in which the structure and function of neurons encounter a progressive loss. This might even lead to the death of neurons (Hardy and Selkoe, 2002). It is characterized by the continuous loss of memory capabilities and several other designated functions. It is believed to be the most common form of age-related cognitive impairment (Blalock et al., 2004; Krishnamurthy et al., 2009). As the principal activity of miRNAs is nothing but to fine-tune the gene functionalities, they have been implicated in many diseases in different ways. Neurodegenerative diseases like AD are no exception to this. There are several pathways defining the diverse panoply of miRNAs connected to AD. Many of the physiological processes underlying AD are influenced by miRNAs in both convergent and divergent approach. This chapter is not aimed at accumulating the list of the miRNAs that are implicated in the AD pathophysiology. For this, the readers should refer to any comprehensive review (Millan, 2017). Instead we concisely touch upon the biological causes behind the progression of AD, and how they are instantiated by the intervention of miRNA activity.

6.1 BACKGROUND OF AD

Let us first explain the basic details about the generation of AD, their causal biogenesis and how miRNAs might be involved in their pathways. This section is a reflection of what we know as the basic causes of AD progression, which is a complex and prominent neurogenerative disorder. The brain is composed of two different types of materials (see Figure 6.1), namely gray matter (GM) and white matter (WM), both of which have designated functions in the normal brain (Millan, 2017). While the former one is a region of AD progression, the latter one is believed to be the region of initiation of AD.

A neuron is a basic unit of the nerve cells (counting about 100 billion), which constitutes a major part of the brain. A neuron is subdivided into the two parts – axon and dendron. The axon of a neuron is connected to the dendron of another neuron. The axon part consists of microtubule bundles. The tau proteins, bound to

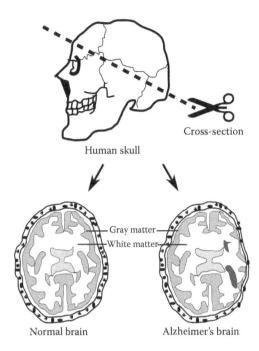

Cross-section

Human skull

Gray matter
White matter

Normal brain   Alzheimer's brain

FIGURE 6.1 Human brain constitutes of GM and WM that experience deformities like cortical shrinkage and inclusion of enlarged ventricles as a result of Alzheimer's progression.

these bundles help the microtubules in the transfer of synaptic information from one neuron to another through the synapse. The normal neuronal connections and their change in the subjects with AD are differentially shown in Figure 6.2. To better understand the principal causes of AD, let us assume that the signaling system in our brain is like a drainage system. As long as the drainage channels (paths through the neurons) are free, the water (signal) can have a smooth flow. Now, there can be two different complications – deposit of garbage in the channel (signal interruption in the channel) or rupture in the drainage channels (breakpoint in channels). In the jargon of AD, the former problem arises due to the accumulation of plaques and the latter one for tangles.

AD is caused by the accumulation of extracellular plaques of amyloidbeta (Aβ) peptides and intracellular neurofibrillary tangles composed of hyperphosphorylated microtubular protein tau (Hauptmann et al., 2006) (see Figure 6.2). Aβ is a naturally occurring, predominantly 40 amino acid long polypeptide (termed as Aβ-40), observed in the cortex region of brain. It is derived from the larger amyloid precursor proteins (APPs). This is an integral membrane protein expressed in many tissues and concentrated in the synapses of neurons. The comprehensive function of APPs is not yet characterized in full. But they have been found to take part in the regulation of synapse formation, neural plasticity and iron export.

In a diseased subject, the tau proteins in axon attain an abnormal form, lose function and detach from the bundles causing microtubule subunits to fall apart

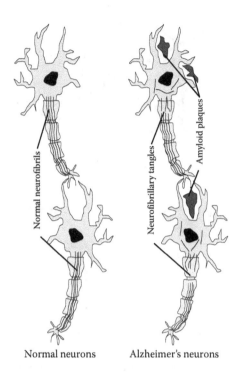

Normal neurons   Alzheimer's neurons

FIGURE 6.2 Formation of tangles within the neuron and plaques outside the neuron (in the extracellular region) due to the accumulation of Aβ-42 in the brain of AD affected subjects.

(Alonso et al., 1996; Kosik et al., 1986). Released tau proteins get clumped together to form tangles. The formation of plaques involves sequential cleavages by two enzymes. Initially, the β-secretase (BACE) enzyme cleaves APP to form soluble extracellular fragment. This is followed by the action of γ-secretase, cleaving within the transmembrane domain. This releases Aβ, which on accumulation forms plaques. In normal brains, α-secretase, another enzyme, cleaves closer to the cell membrane and prevents eventual generation of Aβ peptide. Increase in the proportion of the longer, more neurotoxic form, Aβ-42, results in the formation of higher order aggregates and subsequently, plaque deposition (Carpenter et al., 1993; Galimberti and Scarpini, 2011).

Several experiments showed that the treatment of primary neurons with Aβ-42 evokes a strong change in miRNA profiles with a substantial portion of miRNAs being downregulated, and thus, highlighting the fact that neuronal miRNA expression changes upon exposure to Aβ (Schonrock et al., 2010, 2012). The peptide AB initially causes a downregulation of miRNAs in hippocampal neurons. Notably, the hippocampus is a region of the brain where memories are stored. As plaques are generated from amyloid-beta peptides, they are often termed as amyloid plaques. Thanks to the progress in neuroimaging, amyloid plaques have been characterized into two different types in recent years (Serrano-Pozo et al., 2011). These are termed as dense-core and diffuse plaques. Their differentiable characteristics are highlighted in Table 6.1.

TABLE 6.1

Two Different Types of Plaques and Their Characteristics

	Dense-Core Plaques	Diffuse Plaques
Morphology	Contour	Ill-defined contour
Compact core	Present	Absent
Thioflavin-S staining	Positive	Positive
Congo Red staining	Negative	Negative
Neuritic	Yes (dystrophic neurites)	No
Synaptic loss	Causes	Does not cause
AD pathology	Relevant	Irrelevant

6.2 miRNAs AND AD

The miRNAs intervene into the pathophysiology of AD in many different ways. There are numerous reported results, many of which are, however, in the form of hypothesis. In a broader sense, there are six different areas of interference between miRNAs and AD, namely APP synthesis and processing, inflammatory mechanisms, different stresses and apoptosis, production anomaly of tau, axonal and neuritic disorder, and synaptic dysfunction. The interplay between these activities and the causes of AD are highlighted in Figure 6.3. We describe the activity of miRNAs under each of them hereunder.

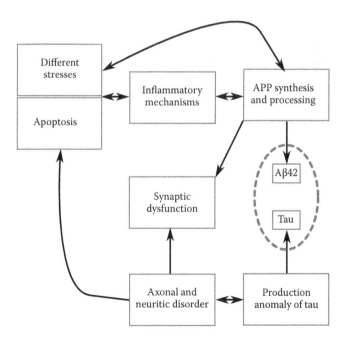

FIGURE 6.3 Interplay between different physiological activities behind the development of AD. There is a strong inter-dependence between a majority of these activities.

6.3 ROLE OF miRNAs

6.3.1 APP SYNTHESIS AND PROCESSING

The aberrant production of Aβ-42 can be mediated by the activity of miRNAs at different levels that include synthesis, alternative splicing, β-secretase processing, γ-secretase processing, and endocytosis and lipid raft incorporation of APP. The APP and its cooperating gene, Amyloid Precursor-Like Protein 2 (Liang et al., 2012), have already been found to be targeted and modified by several miRNAs (Schonrock et al., 2012; Long and Lahiri, 2011; Zhang et al., 2015). The synthesis of APP enhanced by the transcription factor AP-1 due to the stimulation through ERK1/2 (Huang et al., 2017) and its alternative neuronal splicing (by the depletion of endogenous Polypyrimidine Tract-Binding Protein 1 and the resultant increase of Polypyrimidine Tract-Binding Protein 2 (Smith et al., 2011)) are also modulated by a few miRNAs. Some miRNAs have also been found to target Beta-secretase 1 during the β-secretase processing of APP, thereby progressing the AD through Aβ-42 production (Das et al., 2016) and even causing cell death (Fang et al., 2012). This might also involve the inhibition of NF-κB (Shi et al., 2016). The γ-secretase processing of APP is basically modulated through the regulation of Presenilin-1 (Strooper and Chávez-Gutiérrez, 2015), a principle catalytic component of γ-secretase, by the action of miRNAs together with some other cellular components (Frigerio et al., 2013). Other than the above, endocytosis and lipid raft incorporation of APP are also mediated by miRNAs involving the components like BACE1 Ran-binding protein-9 and ATP-Binding Cassette Transporter A1 (Barbato et al., 2014; Akram et al., 2010; Kim et al., 2012).

6.3.2 INFLAMMATORY MECHANISMS

Neuroinflammation is a complex process that overlaps with the pathways of AD by means of strong interactions with immunological mechanisms in the brain involving microglia and astroglia (Heneka et al., 2015). The microglia, a kind of resident phagocytic macrophages of the brain, plays important roles in protecting and restoring the brain from AD progression. It is one of the intersection points through which miRNAs regulate AD (Ager et al., 2010; Asai et al., 2015; Hamelin et al., 2016; Lee and Landreth, 2010). Some miRNAs have recently been observed to cause M2 polarization of microglia (Yu et al., 2017). Different classes of miRNAs either influence the pro-inflammatory mediators like NF-κB (Boldin and Baltimore, 2012; Cui et al., 2010; Li et al., 2011), or IL-1β and Aβ-42 together (Lukiw et al., 2008; Wang et al., 2012). More than this, pro and anti-inflammatory actions of miRNAs that mediate AD have also been identified in recent times (Ponomarev et al., 2011). Extracellular release of miRNAs and their influence on inflammatory processes have also come into notice very recently (Su et al., 2016).

6.3.3 DIFFERENT STRESSES AND APOPTOSIS

Cellular stresses of various kinds and apoptosis have been implicated in AD for a long time. The miRNAs take regulatory part in this in various forms as described hereunder.

6.3.3.1 Oxidative Stress

Oxidative stress has been found to influence the synaptic and mitochondrial dysfunction in the brain, thereby leading to AD (Cencioni et al., 2013; Lu et al., 2014). On the other hand, it is also associated with the hyperphosphorylation of tau proteins (Naini and Soussi-Yanicostas, 2015; Ramamoorthy et al., 2012). miRNAs mediate the oxidative stress mainly by targeting the 15-Lipoxygenase (Zhao et al., 2014). The 15-Lipoxygenase accelerates the accumulation of Aβ-42 by facilitating the production of Beta-secretase 1 via TF Specificity Protein-1 (Chu et al., 2012). In fact, oxidative stress is also linked to the anti-inflammatory activities in microglia (Su et al., 2016). The miRNAs have also been identified to take part, cooperatively with peroxides, in the induction of oxidative stress involving Aβ-42 (Li et al., 2014).

6.3.3.2 Endoplasmic Reticulum (ER) Stress

In a cell, ER makes the newly constructed proteins ready for delivery to the golgi body by folding it properly (Iurlaro and Muñoz-Pinedo, 2016). This mechanism has recently been shown to be interrupted by the production of neurotoxins, presence of cellular stress, and protein overload, thereby drawing a connection with AD. ER stress, associated with the inhibition of miRNAs, has already been found to induce apoptosis (Maurel and Chevet, 2013). Moreover, the miRNA mediated ER stress reduces the activity of Phosphatase and Tensin Homologue (PTEN) that drives AD (Knafo et al., 2016). Recent studies suggest that unfolded protein response regulated by miRNAs is also associated to ER stress.

6.3.3.3 Apoptosis

Apoptosis, which is closely connected to cellular stresses, is a typical feature of AD. It is also related with other important factors like autophagy, mitochondrial and synaptic dysfunctions, etc. Earlier studies have shown that apoptotic cell loss is mediated by miRNAs through the action on Silent Information Regulator 1 (Yamakuchi et al., 2008), Caspase 3, Bcl-2 (Wang et al., 2009), heat shock protein HSPA12B (Chi et al., 2014) and IGF-1 (Fenn et al., 2013; Hu et al., 2013). However, miRNAs also provide some protection mechanisms against the apoptosis in neuronal cells.

6.3.3.4 Mitochondrial Dysfunction

Mitochondrial dysfunction initiates the processes that lead to apoptosis, thereby increasing the chance of AD progression (Hroudová et al., 2014). A few classes of miRNAs have already been reported to have association with the mitochondrial survival factor Bcl-2 (Wang et al., 2009) and the anti-oxidant Superoxide Dismutase 2 (Ji et al., 2013). Some miRNAs are also known to maintain the level of mitochondrial output, highlighting their association with AD in abnormal conditions (Burté et al., 2015; Zhang et al., 2016).

6.3.3.5 Control of Cell Cycle Re-Entry

The cell cycle re-entry is tightly connected with apoptosis, autophagy and neurotoxic protein clearance in the context of miRNA regulation (Ghavami et al., 2014; Mukhopadhyay et al., 2014). The miRNAs might mediate the cell cycle re-entry through the actions on TGFβ-1/TβIIR (Schonrock et al., 2012), E2F1 transcription factor and Retinoblastoma protein (Satoh, 2012), thereby increasing the risk of AD.

6.3.4 PRODUCTION ANOMALY OF TAU

The neurotoxic tau proteins are well connected to the AD pathway. Several miRNAs are known to mediate the production of tau through anomalous synthesis (Iqbal et al., 2016), alternative splicing (Smith et al., 2015), cleavage (Martin et al., 2011), N-acetylglucosamination (Luo et al., 2015), phosphorylation/dephosphorylation (Chang et al., 2014), and acetylation/deacetylation (Liu et al., 2013; Rodriguez-Ortiz et al., 2014; Weinberg et al., 2015). The N-acetyl-glucosamination of tau is known to be controlled with miRNAs by means of targeting the O-GlcNActransferase. The phosphorylation of tau proteins is mainly arbitrated by miRNAs via the action on kinase proteins like ERK-1, CaMKII, Fyn kinase, AMP-activated protein kinase, etc. Post-translational marking of tau also gets controlled by miRNAs, and this is closely connected to anomalous patterns of tau phosphorylation (Banzhaf-Strathmann et al., 2014). On the other side, acetylation of tau takes place through aiming the enzymes p300 and SIRT-1 by miRNAs.

6.3.5 AXONAL AND NEURITIC DISORDER

The disorders in axon and neurites are well connected to the initiation of AD. A majority of the production anomalies of tau proteins are connected to the microtubules that lead to the structural integrity and organization of axons (Pooler et al., 2014; Spillantini and Goedert, 2013). The structural formation, stabilization and elongation of axons and neurites are affected by miRNAs, thereby linking it to AD. The miRNAs keep control over axonal structures by targeting the Neurofilament Heavy Chain (Haramati et al., 2010), axonal stabilization by working on Microtubule-Associated Protein 1B (Dajas-Bailador et al., 2012), axonal elongation by targeting Neurone Navigator 3 (Shioya et al., 2010), and axonal growth by the action on Cytoplasmic Linker Associated Protein-2 (Beffert et al., 2012; Hur et al., 2011). On the other side, miRNAs also control the neuritic extension (Dajas-Bailador et al., 2012) and neuritic growth (Shioya et al., 2010), perturbation of which cause AD. Defective protein trafficking in axons due to the impaired clearance of Aβ-42 and tau have also been implicated in AD by the action of miRNAs in recent years (Sun et al., 2014a,b).

6.3.6 SYNAPTIC DYSFUNCTION

The miRNAs take part in synaptic dysfunction at different levels, namely pre-synaptic, post-synaptic or via dendritic anomalies. Although pre-synaptic regulations are limited in evidence, they have been reported to modulate the activity of synapse-associated proteins like Protein-25 (Wei et al., 2013), Synaptobrevin and Syntaxin-1A (Agostini et al., 2011), and Vesicle-Associated Membrane Protein 2 (Hu et al., 2015). On the other side, miRNAs target the proteins that control homeostasis and plasticity at the post-synaptic level in brain, thus linking it to AD (Garza-Manero et al., 2014). One such protein is Cytoplasmic Linker Associated Protein-2 that is connected to synaptic plasticity and neurotransmission (Beffert et al., 2012). The dendritic anomalies are generally mediated though the action on Actin Related Protein 2/3 Nucleation Complex (Lippi et al., 2011), Cofilin (Gu et al., 2010), and Lim-domain containing Kinase-1 (Bernstein and Bamburg, 2010).

6.4 CONCLUSION

The regulatory involvement of miRNAs has already been established for various disease phenotypes. This chapter explicitly concentrates on the association of this biomolecule with AD. The latest understanding of the mechanism behind Alzheimer's progression and how it associates with the activities of miRNAs are the principal concerns here. We additionally introduce some of the topical hypotheses yet to be critically explored. The miRNAs are not only the tuners of gene activity, but also have greater implications in many diseases. Studying the involvement of miRNAs in AD has explicitly given rise to a number of hypotheses that are yet to be manifested.

One of the interesting questions still to be addressed is the exact involvement of WM in AD and how miRNAs take part in this. A decade-old hypothesis claims that AD may also originate in the WM of brain (Roher et al., 2002). They have in fact shown that alterations in the WM may cause changes in GM. This is in contrast to the previous belief that the GM is primarily affected with AD and then propagates to the WM. This has also been verified later on high-AD-risk groups of women using multimodal imaging (Gold et al., 2010). Similar confirmations have also been received from differential co-expression analyses in recent years (Bhattacharyya and Bandyopadhyay, 2013). It is hard to make a concrete decisions based on these upcoming hypotheses and their preliminary evidence. It might require sex-specific and independent studies to get stronger confirmation. Either we require more specific experimental studies to nullify this initial belief, or otherwise accept this and look more into the reason. Nevertheless, in vitro miRNA activity in WM appears to have some important connection with AD progression.

Another strong belief in the pathophysiology of AD is that the progression of AD is sex-specific. Apparently, women are more susceptible to the development of AD, but one of the reasons could be the higher life expectancy of women. Roher et al. are the first to highlight that WM alterations in AD cases are more predominantly found in women than in men (Roher et al., 2002). This has also been verified in a recent study on high-AD-risk groups of women using multimodal imaging (Gold et al., 2010). Remarkably, the expression profiles of miRNAs that are taken in (Bhattacharyya and Bandyopadhyay, 2013) are also taken from a study on elderly women. Therefore, early Alzheimer's progression in elderly women seems to be associated with some activities of miRNAs (and associated transcription factors and genes) in the WM. A recent survey has argued well the significance of studying sex and gender differences in AD (Carter et al., 2012).

How the response by miRNAs to inflammatory events links to AD progression is another point of interest for deeper analysis (Millan, 2017). It is equally interesting to explore the involvement miRNAs in different cellular phases causing the progression of AD. Profiling of miRNAs in the human brain has recently explored a downregulation of miRNAs, which might aggravate multiple layers of pathology at the molecular and functional level, at the intermediate and late braak stages of AD (Salta and Strooper, 2017). Albeit much of the production anomaly of tau proteins is well characterized, however, the synthesis and splicing activities on tau are not fully understood. These biological phenomena should be tested in greater detail for a better understanding of the interplay between miRNAs and AD. It is further interesting to explore the impact of miRNAs on the epigenetic characteristics that take part in the control of stress stimuli and the mechanisms of ageing pathophysiology related to AD (Cencioni et al., 2013).

ACKNOWLEDGMENT

The work of Malay Bhattacharyya is supported by the Visvesvaraya Young Faculty Research Fellowship 2015-16 of MeitY, Government of India.

REFERENCES

Ager, R. R., M. I. Fonseca, S. Chu et al. (2010). Microglial C5aR (CD88) expression correlates with amyloid-β deposition in murine models of alzheimer's disease. *Journal of Neurochemistry* 113 (2): 389–401.

Agostini, M., P. Tucci, R. Killick, E. et al. (2011). Neuronal differentiation by TAp73 is mediated by microRNA-34a regulation of synaptic protein targets. *Proceedings of the National Academy of Sciences USA* 108 (52): 21093–21098.

Akram, A., J. Schmeidler, P. Katsel et al. (2010). Increased expression of cholesterol transporter ABCA1 is highly correlated with severity of dementia in AD hippocampus. *Brain Research* 1318: 167–177.

Alonso, A. C., I. Grundke-Iqbal, and K. Iqbal (1996). Alzheimer's disease hyperphosphorylated tau sequesters normal tau into tangles of filaments and disassembles microtubules. *Nature Medicine* 2 (7): 783–787.

Asai, H., S. Ikezu, S. Tsunoda et al. (2015). Depletion of microglia and inhibition of exosome synthesis halt tau propagation. *Nature Neuroscience* 18 (11): 1584–1593.

Bandyopadhyay, S. and M. Bhattacharyya (2009). Analyzing miRNA coexpression networks to explore TF-miRNA regulation. *BMC Bioinformatics* 10: 163.

Banzhaf-Strathmann, J., E. Benito, S. May et al. (2014). MicroRNA-125b induces tau hyperphosphorylation and cognitive deficits in Alzheimer's disease. *The EMBO Journal*, e201387576.

Barbato, C., S. Pezzola, C. Caggiano et al. (2014). A lentiviral sponge for miR-101 regulates RanBP9 expression and amyloid precursor protein metabolism in hippocampal neurons. *Frontiers in Cellular Neuroscience* 8: 37.

Bartel, D. P. (2004). MicroRNAs: Genomics, biogenesis, mechanism, and function. *Cell* 116 (2): 281–297.

Bartel, D. P. (2009). MicroRNAs: Target recognition and regulatory functions. *Cell* 136 (2): 215–233.

Beffert, U., G. M. Dillon, J. M. Sullivan et al. (2012). Microtubule plus-end tracking protein CLASP2 regulates neuronal polarity and synaptic function. *Journal of Neuroscience* 32 (40): 13906–13916.

Bernstein, B. W. and J. R. Bamburg (2010). ADF/cofilin: A functional node in cell biology. *Trends in Cell Biology* 20 (4): 187–195.

Bhattacharyya, M. and S. Bandyopadhyay (2013). Studying the differential co-expression of microRNAs reveals significant role of white matter in early Alzheimer's progression. *Molecular Biosystems* 9: 457–466.

Bhattacharyya, M., L. Feuerbach, T. Bhadra et al. (2012). MicroRNA transcription start site prediction with multi-objective feature selection. *Statistical Applications in Genetics and Molecular Biology* **11** (1): 6.

Blalock, E. M., J. W. Geddes, K. C. Chen et al. (2004). Incipient Alzheimer's disease: Microarray correlation analyses reveal major transcriptional and tumor suppressor responses. *Proceedings of the National Academy of Sciences USA* **101** (7): 2173–2178.

Boldin, M. P. and D. Baltimore (2012). MicroRNAs, new effectors and regulators of NF-κB. *Immunological Reviews* **246** (1): 205–220.

Burté, F., V. Carelli, P. F. Chinnery et al. (2015). Disturbed mitochondrial dynamics and neurodegenerative disorders. *Nature Reviews Neurology* **11** (1): 11–24.

Carpenter, M. K., K. A. Crutcher, and S. B. Kater (1993). An analysis of the effects of Alzheimer's plaques on living neurons. *Neurobiology of Aging* **14** (3): 207–215.

Carter, C. L., E. M. Resnick, M. Mallampalli et al. (2012). Sex and gender differences in Alzheimer's disease: Recommendations for future research. *Journal of Women's Health* **21** (10): 1–6.

Cencioni, C., F. Spallotta, F. Martelli et al. (2013). Oxidative stress and epigenetic regulation in ageing and age-related diseases. *International Journal of Molecular Sciences* **14** (9): 17643–17663.

Chang, F., L. Zhang, W. Xu et al. (2014). microRNA-9 attenuates amyloidβ-induced synaptotoxicity by targeting calcium/calmodulin-dependent protein kinase kinase 2. *Molecular Medicine Reports* **9** (5): 1917–1922.

Chen, C. Z. (2005). MicroRNAs as oncogenes and tumor suppressors. *New England Journal of Medicine* **353** (17): 1765–1771.

Chi, W., F. Meng, Y. Li et al. (2014). Impact of microRNA-134 on neural cell survival against ischemic injury in primary cultured neuronal cells and mouse brain with ischemic stroke by targeting HSPA12B. *Brain Research* **1592**: 22–33.

Chu, J., J. Zhuo, and D. Praticò (2012). Transcriptional regulation of βsecretase-1 by 12/15-lipoxygenase results in enhanced amyloidogenesis and cognitive impairments. *Annals of Neurology* **71** (1): 57–67.

Corcoran, D. L., K. V. Pandit, B. Gordon et al.(2009). Features of mammalian microRNA promoters emerge from polymerase II chromatin immunoprecipitation data. *PLoS ONE* **4** (4): e5279.

Cui, J. G., Y. Y. Li, Y. Zhao et al.(2010). Differential regulation of interleukin-1 receptor-associated kinase-1 (IRAK-1) and IRAK-2 by microRNA-146a and NF-κB in stressed human astroglial cells and in Alzheimer disease. *Journal of Biological Chemistry* **285** (50): 38951–38960.

Dajas-Bailador, F., B. Bonev, P. Garcez et al. (2012). microRNA-9 regulates axon extension and branching by targeting Map1b in mouse cortical neurons. *Nature Neuroscience* **15** (5): 697–699.

Das, U., L. Wang, A. Ganguly et al. (2016). Visualizing APP and BACE-1 approximation in neurons yields insight into the amyloidogenic pathway. *Nature Neuroscience* **19** (1): 55–64.

Fang, M., J. Wang, X. Zhang et al. (2012). The miR-124 regulates the expression of BACE1/β-secretase correlated with cell death in Alzheimer's disease. *Toxicology Letters* **209** (1): 94–105.

Fenn, A. M., K. M. Smith, A. E. Lovett-Racke et al. (2013). Increased micro-RNA 29b in the aged brain correlates with the reduction of insulin-like growth factor-1 and fractalkine ligand. *Neurobiology of Aging* **34** (12): 2748–2758.

Frigerio, C. S., P. Lau, E. Salta et al. (2013). Reduced expression of hsa-miR-27a-3p in CSF of patients with Alzheimer disease. *Neurology* **81** (24): 2103–2106.

Galimberti, D. and E. Scarpini (2011). Alzheimer's disease: From pathogenesis to disease-modifying approaches. *CNS & Neurological Disorders—Drug Targets* **10** (2): 163–174.

Garza-Manero, S., I. Pichardo-Casas, C. Arias et al. (2014). Selective distribution and dynamic modulation of miRNAs in the synapse and its possible role in Alzheimer's disease. *Brain Research* **1584**: 80–93.

Ghavami, S., S. Shojaei, B. Yeganeh et al. (2014). Autophagy and apoptosis dysfunction in neurodegenerative disorders. *Progress in Neurobiology* **112**: 24–49.

Gold, B. T., D. K. Powell, A. H. Andersen et al. (2010). Alterations in multiple measures of white matter integrity in normal women at high risk for Alzheimer's disease. *NeuroImage* **52** (4): 1487–1494.

Gu, J., C. W. Lee, Y. Fan et al. (2010). ADF/cofilin-mediated actin dynamics regulate AMPA receptor trafficking during synaptic plasticity. *Nature Neuroscience* **13** (10): 1208–1215.

Hamelin, L., J. Lagarde, G. Dorothée et al. (2016). Early and protective microglial activation in Alzheimer's disease: A prospective study using 18F-DPA-714 PET imaging. *Brain* **139** (4): 1252–1264.

Haramati, S., E. Chapnik, Y. Sztainberg et al. (2010). miRNA malfunction causes spinal motor neuron disease. *Proceedings of the National Academy of Sciences USA* **107** (29): 13111–13116.

Hardy, J. and J. D. Selkoe (2002). The amyloid hypothesis of Alzheimer's disease: Progress and problems on the road to therapeutics. *Science* **297** (5580): 353–356.

Hauptmann, S., U. Keil, I. Scherping et al. (2006). Mitochondrial dysfunction in sporadic and genetic Alzheimer's disease. *Experimental Gerontology* **41** (7): 668–673.

Heneka, M. T., M. J. Carson, J. E. Khoury et al. (2015). Neuroinflammation in Alzheimer's disease. *The Lancet Neurology* **14** (4): 388–405.

Hroudová, J., N. Singh, and Z. Fišar (2014). Mitochondrial dysfunctions in neurodegenerative diseases: Relevance to Alzheimer's disease. *BioMed Research International 2014* (175062).

Hu, S., H. Wang, K. Chen et al. (2015). MicroRNA-34c downregulation ameliorates amyloid-β-induced synaptic failure and memory deficits by targeting VAMP2. *Journal of Alzheimer's Disease* **48** (3): 673–686.

Hu, Y., X. Wang, L. Li et al. (2013). MicroRNA-98 induces an Alzheimer's disease-like disturbance by targeting insulin-like growth factor 1. *Neuroscience Bulletin* **29** (6), 745–751.

Huang, Y. A., B. Zhou, M. Wernig et al. (2017). ApoE2, ApoE3, and ApoE4 differentially stimulate APP transcription and Aβ secretion. *Cell* **168** (3): 427–441.

Hur, E., B. D. Lee, S. Kim et al. (2011). GSK3 controls axon growth via CLASP-mediated regulation of growth cone microtubules. *Genes & Development* **25** (18): 1968–1981.

Iqbal, K., F. Liu, and C. Gong (2016). Tau and neurodegenerative disease: The story so far. *Nature Reviews Neurology* **12** (1): 15–27.

Iurlaro, R. and C. Muñoz-Pinedo (2016). Cell death induced by endoplasmic reticulum stress. *FEBS Journal* **283** (14): 2640–2652.

Ji, G., K. Lv, H. Chen et al. (2013). MiR-146a Regulates SOD2 Expression in H2O2 Stimulated PC12 Cells. *PloS One* **8** (7): e69351.

Kim, J., H. Yoon, C. M. Ramírez et al. (2012). miR-106b impairs cholesterol efflux and increases Aβ levels by repressing ABCA1 expression. *Experimental Neurology* **235** (2): 476–483.

Kim, V. N. (2005). MicroRNA biogenesis: Coordinated cropping and dicing. *Nature Reviews Molecular Cell Biology* **6** (5): 376–385.

Knafo, S., C. Sánchez-Puelles, E. Palomer et al. (2016). PTEN recruitment controls synaptic and cognitive function in Alzheimer's models. *Nature Neuroscience* **19** (3): 443–453.

Kosik, K. S., C. L. Joachim, and D. J. Selkoe (1986). Microtubule-associated protein tau (τ) is a major antigenic component of paired helical filaments in Alzheimer disease. *Proceedings of the National Academy of Sciences USA* **83** (11): 4044–4048.

Krishnamurthy, V., N. S. Issac, and J. Natarajan (2009). Computational identification of Alzheimer's disease specific transcription factors using microarray gene expression data. *Journal of Proteomics & Bioinformatics* **2** (12): 505–508.

Lauressergues, D., J. Couzigou, H. S. Clemente et al. (2015). Primary transcripts of microRNAs encode regulatory peptides. *Nature* **520**: 90–93.

Lee, C. Y. D. and G. E. Landreth (2010). The role of microglia in amyloid clearance from the AD brain. *Journal of Neural Transmission* **117** (8): 949–960.

Li, J. J., G. Dolios, R. Wang et al. (2014). Soluble beta-amyloid peptides, but not insoluble fibrils, have specific effect on neuronal microRNA expression. *PloS One* **9** (3): e90770.

Li, Y. Y., J. G. Cui, P. Dua et al. (2011). Differential expression of miRNA-146a-regulated inflammatory genes in human primary neural, astroglial and microglial cells. *Neuroscience Letters* **499** (2): 109–113.

Liang, C., H. Zhu, Y. Xu et al. (2012). MicroRNA-153 negatively regulates the expression of amyloid precursor protein and amyloid precursor-like protein 2. *Brain Research* **1455**: 103–113.

Lippi, G., J. R. Steinert, E. L. Marczylo et al. (2011). Targeting of the Arpc3 actin nucleation factor by miR-29a/b regulates dendritic spine morphology. *The Journal of Cell Biology* **194** (6): 889–904.

Liu, C., R. Wang, Z. Jiao et al. (2013). MicroRNA-138 and SIRT1 form a mutual negative feedback loop to regulate mammalian axon regeneration. *Genes & Development* **27** (13): 1473–1483.

Long, J. and D. Lahiri (2011). Current drug targets for modulating Alzheimer's amyloid precursor protein: Role of specific micro-RNA species. *Current Medicinal Chemistry* **18** (22): 3314–3321.

Lu, T., L. Aron, J. Zullo et al. (2014). REST and stress resistance in ageing and Alzheimer's disease. *Nature* **507** (7493): 448–454.

Lukiw, W. J., Y. Zhao, and J. G. Cui (2008). An NF-κB-sensitive microRNA-146a-mediated inflammatory circuit in Alzheimer disease and in stressed human brain cells. *Journal of Biological Chemistry* **283** (46): 31315–31322.

Luo, P., T. He, R. Jiang et al. (2015). MicroRNA-423-5p targets O-GlcNAc transferase to induce apoptosis in cardiomyocytes. *Molecular Medicine Reports* **12** (1): 1163–1168.

Martin, L., X. Latypova, and F. Terro (2011). Post-translational modifications of tau protein: Implications for Alzheimer's disease. *Neurochemistry International* **58**: 458–471.

Maurel, M. and E. Chevet (2013). Endoplasmic reticulum stress signaling: The microRNA connection. *American Journal of Physiology—Cell Physiology* **304** (12): C1117–C1126.

Millan, M. J. (2017). Linking deregulation of non-coding RNA to the core pathophysiology of Alzheimer's disease: An integrative review. *Progress in Neurobiology*. **156**: 1–68.

Mukhopadhyay, S., P. K. Panda, N. Sinha et al. (2014). Autophagy and apoptosis: Where do they meet? *Apoptosis* **19** (4): 555–566.

Naini, S. M. A. and N. Soussi-Yanicostas (2015). Tau hyperphosphorylation and oxidative stress, a critical vicious circle in neurodegenerative tauopathies? *Oxidative Medicine and Cellular Longevity* **2015** (151979).

Ponomarev, E. D., T. Veremeyko, N. Barteneva et al. (2011). MicroRNA-124 promotes microglia quiescence and suppresses EAE by deactivating macrophages via the C/EBP-α-PU. 1 pathway. *Nature Medicine* **17** (1): 64–70.

Pooler, A. M., W. Noble, and D. P. Hanger (2014). A role for tau at the synapse in Alzheimer's disease pathogenesis. *Neuropharmacology* **76**: 1–8.

Ramamoorthy, M., P. Sykora, M. Scheibye-Knudsen et al. (2012). Sporadic Alzheimer disease fibroblasts display an oxidative stress phenotype. *Free Radical Biology and Medicine* **53** (6): 1371–1380.

Rodriguez-Ortiz, C. J., D. Baglietto-Vargas, H. Martinez-Coria et al. (2014). Upregulation of miR-181 decreases c-Fos and SIRT-1 in the hippocampus of 3xTg-AD mice. *Journal of Alzheimer's Disease* **42** (4): 1229–1238.

Roher, A. E., N. Weiss, T. A. Kokjohn et al. (2002). Increased Aβ peptides and reduced cholesterol and myelin proteins characterize white matter degeneration in alzheimer's disease. *Biochemistry* **41** (37): 11080–11090.

Salta, E. and B. D. Strooper (2017). microRNA-132: A key noncoding RNA operating in the cellular phase of Alzheimer's disease. *The FASEB Journal* **31** (2): 424–433.

Satoh, J. (2012). Molecular network of microRNA targets in Alzheimer's disease brains. *Experimental Neurology* **235** (2): 436–446.

Schonrock, N., Y. D. Ke, D. Humphreys et al. (2010). Neuronal microRNA deregulation in response to Alzheimer's disease amyloid-β. *PLoS ONE* **5** (6): e11070.

Schonrock, N., M. Matamales, L. M. Ittner et al. (2012). MicroRNA networks surrounding APP and amyloid-β metabolism – implications for Alzheimer's disease. *Experimental Neurology* **235** (2): 447–454.

Serrano-Pozo, A., M. P. Frosch, E. Masliah et al. (2011). Neuropathological alterations in Alzheimer disease. *Cold Spring Harbor Perspectives in Medicine* **1** (1): a006189.

Shi, Z.-M., Y.-W. Han, X.-H. Han et al. (2016). Upstream regulators and downstream effectors of NF-κB in Alzheimer's disease. *Journal of the Neurological Sciences* **366**:127–134.

Shioya, M., S. Obayashi, H. Tabunoki et al. (2010). Aberrant microRNA expression in the brains of neurodegenerative diseases: miR-29a decreased in Alzheimer disease brains targets neurone navigator 3. *Neuropathology and Applied Neurobiology* **36** (4): 320–330.

Smith, P., A. A. Hashimi, J. Girard et al. (2011). In vivo regulation of amyloid precursor protein neuronal splicing by microRNAs. *Journal of Neurochemistry* **116** (2): 240–247.

Smith, P. Y., J. Hernandez-Rapp, F. Jolivette et al. (2015). miR-132/212 deficiency impairs tau metabolism and promotes pathological aggregation in vivo. *Human Molecular Genetics* **24** (23): 6721–6735.

Spillantini, M. G. and M. Goedert (2013). Tau pathology and neurodegeneration. *The Lancet Neurology* **12** (6): 609–622.

Strooper, B. D. and L. Chávez-Gutiérrez (2015). Learning by failing: Ideas and concepts to tackle γ-secretases in Alzheimer's disease and beyond. *Annual Review of Pharmacology and Toxicology* **55**: 419–437.

Su, W., M. S. Aloi, and G. A. Garden (2016). MicroRNAs mediating CNS inflammation: Small regulators with powerful potential. *Brain, Behavior, and Immunity* **52**: 1–8.

Sun, X., Y. Wu, M. Gu et al. (2014a). Selective filtering defect at the axon initial segment in Alzheimer's disease mouse models. *Proceedings of the National Academy of Sciences USA* **111** (39): 14271–14276.

Sun, X., Y. Wu, M. Gu et al. (2014b). miR-342-5p decreases ankyrin G levels in Alzheimer's disease transgenic mouse models. *Cell Reports* **6** (2): 264–270.

Trujillo, R. D., S. Yue, Y. Tang et al. (2010). The potential functions of primary microRNAs in target recognition and repression. *EMBO Journal* **29**: 3272–3285.

Wang, L., Y. Huang, G. Wang et al. (2012). The potential role of microRNA-146 in Alzheimer's disease: Biomarker or therapeutic target? *Medical Hypotheses* **78** (3): 398–401.

Wang, X., P. Liu, H. Zhu et al. (2009). miR-34a, a microRNA up-regulated in a double transgenic mouse model of Alzheimer's disease, inhibits bcl2 translation. *Brain Research Bulletin* **80** (4): 268–273.

Wei, C., E. J. Thatcher, A. F. Olena et al. (2013). miR-153 regulates SNAP-25, synaptic transmission, and neuronal development. *PLoS One* **8** (2): e57080.

Weinberg, R. B., E. J. Mufson, and S. E. Counts (2015). Evidence for a neuroprotective microRNA pathway in amnestic mild cognitive impairment. *Frontiers in Neuroscience* **9**: 430.

Yamakuchi, M., M. Ferlito, and C. J. Lowenstein (2008). miR-34a repression of SIRT1 regulates apoptosis. *Proceedings of the National Academy of Sciences USA* **105** (36): 13421–13426.

Yu, A., T. Zhang, H. Duan et al. (2017). MiR-124 contributes to M2 polarization of microglia and confers brain inflammatory protection via the C/EBP-α pathway in intracerebral hemorrhage. *Immunology Letters* **182**: 1–11.

Zhang, B., C. F. Chen, A. H. Wang et al. (2015). MiR-16 regulates cell death in alzheimer's disease by targeting amyloid precursor protein. *European Review for Medical and Pharmacological Sciences* **19** (21): 4020–4027.

Zhang, R., H. Zhou, L. Jiang et al. (2016). MiR-195 dependent roles of mitofusin2 in the mitochondrial dysfunction of hippocampal neurons in SAMP8 mice. *Brain Research* **1652**: 135–143.

Zhao, Y., S. Bhattacharjee, B. M. Jones et al. (2014). Regulation of neurotropic signaling by the inducible, NF-kB-sensitive miRNA-125b in Alzheimer's disease (AD) and in primary human neuronal-glial (HNG) cells. *Molecular Neurobiology* **50** (1): 97–106.

7 MicroRNAs in Neurogenesis and Neurodegeneration

Mayuresh Anant Sarangdhar and Beena Pillai

CONTENTS

Introduction

Neurogenesis is the process by which neurons are born, shape up, and connect to other neurons in the brain. By contrast, neurodegeneration refers to the loss of neurons following physical injury, chemical insults, pathological neuronal cell death due to mutations and infections, or natural processes such as aging. Neurogenesis happens in massive waves during a small window of embryonic development when stem cells divide rapidly, increasing the numbers of progenitor cells, which further differentiate into immature neurons and, finally, under the influence of many extracellular cues, mature into individual neurons. It is now believed that no two neurons may be absolutely identical. Some of them may undertake a long migration through the developing brain, eventually docking into their place of action, while others may grow axons that traverse a dense forest of cells to connect up with distant targets, pulsing electrical signals through highways of communication. These connections are far from rigid joints on a set path; they are highly dynamic, growing, reshaping, pruning away unused connections, and strengthening the points where messages are transmitted with regularity. The distinctive features of the human race—the ability to imitate, learn, memorize, empathize, communicate, and cooperate—may indeed depend on the way our neurons connect up.

Although the adult brain does retain the potential to produce new neurons and rewire some existing connections, this capacity is muted with age. Instead, some neurons are lost, sometimes abruptly, following accidents or stroke, and more often gradually, through the irreversible process of aging. Environmental factors such as diet and exercise and genetic factors such as mutations may expedite or prolong this process. The neurons that are lost are hardly replaced, thus leading to some blocks and barricades in the information highways. Organs such as the skin, blood, and liver can regenerate rapidly after injury, a quality that makes live organ donation and graft possible. Engineering has provided solutions for repairing bones or the heart through the transplant of titanium rods or pacemakers. Although, the brain or peripheral nerves can be neither readily grafted nor replaced by engineered devices yet, it is now clear that the adult brain and nerves do have the latent ability to regenerate, an ability that is readily seen in lower organisms that can regenerate severed nerves such as the axolotl and earthworm. Pharmacological augmentation of this ability is indeed one of the promising avenues for repair of nerves and the brain.

A thorough understanding of the way our neurons are born and their numbers are regulated is, therefore, essential. Further, factors that can direct the migration and proper integration of newborn neurons to their targets along with the ability to control the process of neurogenesis are all necessary. Besides neurons, the nervous system consists of other cell types that are equally important—the glial cells, composed of astrocytes, oligodendrocytes, and the developmentally distinct microglia. The glia originates from the same stem cells as neurons do, but in a controlled cascade of events that occur in a strict temporal sequence of neurogenesis followed by gliogenesis. The factors that control these important developmental transition points from neurogenesis to gliogenesis are also, therefore, key players in neurogenesis. These developmental windows of time, when neural stem cells proliferate rapidly, do not occur in all parts of the brain simultaneously. Cortical neurons may already be maturing when the hippocampal neurons are born, suggesting that spatial control of signaling molecules and the developmental programs they set in motion are important in orchestrating neurogenesis in a spatiotemporally regulated manner.

At the heart of this spatiotemporally restricted program of neurogenesis are molecular pathways that control cell numbers, rates of proliferation, and the morphology and functionality of the neurons and glia in the brain. The brain is unique in its ability to express the largest diversity of genes, a reflection of the cellular complexity necessary to carry out an overwhelming inventory of tasks. In many gene families, the brain may express a distinct ortholog; for instance, the aging brain may replace as much as 96% of an otherwise ubiquitous, core histone protein with a histone variant. It also adopts extensive alternative splicing, significant levels of RNA editing and unusual modes of metabolic regulation, all of which require specialized regulatory control.

The gene regulatory mechanisms of the brain require the concerted action of chromatin organizers, transcription factors and noncoding RNAs. Noncoding RNAs include several classes: long noncoding RNAs (lncRNAs), microRNAs (miRNAs), and piwi-interacting RNAs, to name a few, are an integral part of gene regulatory module. Although the nomenclature overtly refers to size differences, they also differ in the proteins they interact with and the functions they perform. LncRNAs have the capacity, through structural motifs and interacting partners, to act as scaffolds for ribonucleoprotein complexes that may form molecular machines in themselves. They can provide tandem-binding sites to sponge off other RNAs or sequester proteins, a function that may be greatly augmented by the formation of covalently closed circular RNAs that are resistant to enzymatic attack. They may also direct modifications of DNA, silencing certain genes (Iyengar et al. 2014). In comparison with lncRNAs, miRNAs have a limited repertoire of functionalities, doubtless owing to their diminutive size and the structural limitations it poses. Offsetting these limitations are certain features of miRNAs that account for their rapid rise as candidates for development of diagnostics and therapeutics. MiRNAs have a highly conserved mode of action—they form the gene-specific core of enzymatic machinery. Thus, the consequences of manipulating miRNAs are relatively more predictable. They interact with targets through set rules of RNA–RNA interaction. Although there are many ways in which a target may escape binding by miRNAs, we can, to a large extent, avoid complications of charge- and shape-based three-dimensional interactions in the selection of miRNAs to target specific genes. Lastly, their small size and stability means that they can be delivered *in vivo*, inside target cells in complex with lipids, peptides, and nanoparticles. Exosomes are naturally occurring extracellular vesicles that may deliver miRNAs, along with other molecular cargo, through body fluids, establishing an interorgan communication channel.

Several observations point to the fact that miRNAs form an important component of neural gene regulatory networks. Firstly, the central and peripheral nervous tissue expresses several miRNAs that are specific to neural cell types. In other words, there are specific neuronal and glial miRNAs that are not expressed elsewhere in the body. The dysregulation of these miRNAs are reported in a variety of neurodegenerative diseases, implying their critical role in sustaining the proper functioning of neural cells. The expression of neuronal or glial miRNAs may be established during development and sustained through reinforcing regulatory pathways. Yet other miRNAs may be rapidly accumulated or degraded following cellular perturbations, thus becoming part of the dynamic regulatory networks. Secondly, many genes known to be important for brain development, either through *in vitro* models or because they are sites for disease-causing mutations, carry miRNA-binding sites in their transcripts. Perturbations of these miRNAs can thus lead to rewiring of gene regulatory pathways, reestablishing cellular fates and roles. Neural cells have unique morphologies that underlie their functional roles. Interneurons may be bipolar to accommodate their role of connecting two distant neurons, while Purkinje neurons are unipolar and motor neurons may have long axons. Some intracellular sites, although connected by the same cytoplasm, may be several microns away in a distinct subcellular compartment. The synapse presents such an intracellular niche with its distinct set of resident proteins, neurotransmitter vesicles, and cellular signaling events. Some miRNAs show localization to such distinct intracellular sites, while others can reshape dendrites and axon termini, thus mediating extranuclear, local gene regulatory modules.

This chapter presents an overview of miRNA action through various stages of neurogenesis. Through their ability to regulate the cell cycle, several miRNAs regulate neural cell proliferation. Later, as these cells differentiate into mature neurons, miRNAs make sure that they restrict the cells to certain fates, by checking the expression of alternative differentiation programs. Yet another swathe of miRNAs directs the migration of neurons or renders the shape of these neurons, through the continuing process of axonogenesis and dendritogenesis. The last class of neuronal miRNAs to be discussed here are the ones that regulate normal functions of mature neurons. This chapter also discusses the consequence of loss or dysregulation of these miRNAs, thus highlighting their involvement in neurodegeneration.

There are thousands of miRNAs in the mammalian genome, some of them primate specific, while the vast majority are highly conserved among vertebrate. This chapter focuses on selected miRNAs that have been studied extensively or have had tremendous, early impact on our understanding of a phenomenon. Some are chosen because of their prototypical nature; others for their unique or exceptional roles. This account is, therefore, not a comprehensive catalog of neural miRNAs, for which the reader is directed to excellent reviews (Table 7.1) (Fineberg, Kosik, and Davidson 2009; Kosik 2006; McNeill and Van Vactor 2012), but is an attempt to present the diverse mechanisms by which miRNAs direct key steps of neurogenesis. This chapter also attempts to integrate the current knowledge of miRNA-mediated gene regulatory mechanisms with some daunting questions and promises of their utility in the treatment of neuropsychiatric diseases.

7.1   MicroRNAs IN NEURAL CELL PROLIFERATION

Controlled neural stem cell proliferation is essential to generate the appropriate number of cells during fetal brain development as well as after injury or stress in adults. An altered number of cell division cycles of neural stem cells may create variation in the cell number, which affects the functions of neural circuits drastically. The mouse brain has been used extensively as a model for studying cellular and molecular factors that govern neural stem cell proliferation and differentiation and formation of functional neural circuits. The mouse embryonic brain development spans from about the 10th or 11th day of gestation (referred to as E10 and E11, respectively) and lasts up to days after birth roughly coinciding with the 3–32 weeks of gestation in humans (Figure 7.1).

In the brain, cortical neurogenesis has been studied extensively and is presented here as a model for understanding the activities of neural stem cells. Initially, neural stem cells proliferate in a pseudostratified epithelial layer, increasing their number. This layer marked by the expression of genes such as Nestin, Sox2, and BMI-1 can be readily visualized through microscopy in transgenic animals carrying fluorescent marker genes under the control of specific promoters. Neuroepithelial cells have an elongated shape with cytoplasmic connections to the apical as well as the basal layer. These cells proliferate and elongate to form radial glial cells, which divide either symmetrically or asymmetrically. Asymmetric division gives rise to basal progenitor cells that migrate outward along the long processes of radial glial cells. The migrating basal progenitors also proliferate to form more basal progenitors or differentiate into mature neurons that finally occupy the outer layers of the brain. By the time the pups are born, the neurons are forming axons and dendrites and subsequently synaptic connections. Postnatal days are marked by the switch to gliogenesis, during which the same stem cells give rise to glial progenitors. In other words, if the cell division of a stem cell

TABLE 7.1
Summary of miRNAs Involved in Key Processes during Neurogenesis

Dendritic Spine Morphogenesis	Neuron Migration	Neuron Maturation	Neural Stem Cell Differentiation	Basal Progenitors or Intermediate Progenitor Proliferation	Neural Stem Cell Proliferation	miRNA/ Developmental Stage
			TRIM32 (Schwamborn, Berezikov, and Knoblich 2009)			Let7
						Let7b
			Tlx.cyclin D1 (Zhao et al. 2010)	Ak1 or p21 (Pollock et al. 2014)		miR-7
REST (Giusti et al. 2014)	Stathmin (Delaloy et al. 2017)		Foxg1 (Shibata et al. 2008)		Stathmin (Delaloy et al. 2017)	miR-9
	Rapgef2 (Han et al. 2016)		TET3 (Lv et al. 2014)			miR-15b
	CoREST (Volvert et al. 2017)				PTEN (Bian et al. 2017)	miR-19
		DCX (Mollinari et al. 2015)	ICAT (Shin et al. 2014)	Max (Berenguer et al. 2013)		miR-22
					DCX (Mollinari et al. 2015)	miR-29b
			Syt1 and Atg9a (Morgado et al. 2015)			miR-34a
					Tbr2 (Bian et al. 2017)	miR-92
	CoREST (Volvert et al. 2017)		(1) Ezh2 (Neo et al. 2014); (2) SCP1 (Visvanathan et al. 2007); (3) PTBP1 (Makeyev et al. 2007); (4) SP1 (Santos et al. 2016)			miR-124

(Continued)

TABLE 7.1 (CONTINUED)
Summary of miRNAs Involved in Key Processes during Neurogenesis

Dendritic Spine Morphogenesis	Neuron Migration	Neuron Maturation	Neural Stem Cell Differentiation	Basal Progenitors or Intermediate Progenitor Proliferation	Neural Stem Cell Proliferation	miRNA/ Developmental Stage
NR2A (Edbauer et al. 2010)			TBC1D1, SGPL1, DGAT1 (Le et al. 2009)			miR-125
	Phf6 (Franzoni et al. 2015)		PCM1 (Zhang et al. 2016)			miR-128
(Magill et al. 2010)			SP1 (Santos et al. 2016)			miR-132
		Mib1 (Smrt et al. 2010)	SP1 (Santos et al. 2016)			miR-137
APT1 (Siegel et al. 2009)						miR-138
					MBD1 (Liu et al. 2010)	miR-184
					Hes1 (Shi et al. 2015)	miR-381
	N-cadherin (Rago et al. 2014)		N-cadherin (Rago et al. 2014)			miR-369-3p
	N-cadherin (Rago et al. 2014)		N-cadherin (Rago et al. 2014)			miR-496
	N-cadherin (Rago et al. 2014)		N-cadherin (Rago et al. 2014)			miR-543

Note: References to the original reports are in brackets. The intersecting cell features the target gene, targeted by a miRNA (row) to affect the process (column).

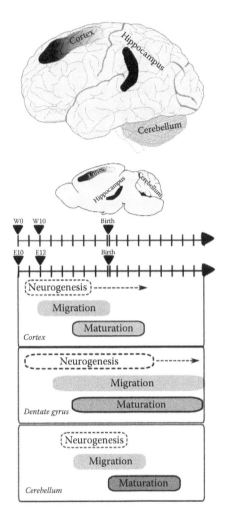

FIGURE 7.1 Comparison of the developmental time lines of human and mouse cortices, dentate gyri (part of hippocampus), and cerebellums. The dotted arrows indicate a slow rate of neurogenesis. Human brain development is presented in weeks (W), while mouse brain development is presented in terms of the embryonic age (E).

produces two identical daughters, then both are expected to continue in the same way as the stem cells (Gotz and Huttner 2005; Martynoga, Drechsel, and Guillemot 2012).

Several miRNAs can keep neural stem cells in a proliferative state, efficiently promoting the production of more neural stem cells. For instance, miR-381 regulates Hes1 and leads to increased Nestin expression and increased number of neurons, but not glia (Shi et al. 2015). Another miRNA, which is enriched in the neural progenitors and regulates their proliferation, is miR-7. Transgenic mice expressing miR-7 sponges—transcripts with multiple miR-7-binding sites to "sponge away" or sequester the endogenous miR-7—developed smaller brains with fewer neuronal cells. Although the number of radial glia cells (RGCs) (neural stem cells) was not affected, dysregulation of the miR-7 target Ak1 in these mice resulted in intermediate progenitor cells undergoing apoptotic cell death (Pollock et al. 2014).

An intriguing feature of neurogenesis is that a good number of cells produced by a wave of rapid proliferation are culled through a phase of massive apoptosis. Downregulation of miR-29b is a critical event in triggering this apoptotic sculpting of the developing brain, since it simultaneously targets a number of transcripts that code for multiple, redundant BH3-only proteins. Subsequently, the increase in expression of miR-29b keeps the apoptotic pathway in check (Kole et al. 2011). Although the fundamental need for a phase of apoptosis in shaping the brain is not clear to us, the molecular challenge of coordinate regulation is elegantly executed by a miRNA.

7.1.1 ASYMMETRIC CELL DIVISION

Asymmetric distribution of cellular factors, usually due to the plane along which cell division occurs, ends up creating unequal daughter cells. In the neural stem cell niche, the plane of division results in the formation of Numb-positive daughter cells that block Notch signaling and take up the alternative, neuronal fate (Zhong et al. 2017). The cells that are destined to make neurons exit the cell cycle and start their outward migration, but the ones that continue to proliferate remain in the niche of the ventricular zone, a region favorable for neurogenesis. The choice between symmetric versus asymmetric cell division by neural progenitors is a major determinant of brain size and function, as premature termination of the proliferative state by inducing asymmetric divisions leads to a smaller brain size and altered balance of different brain cell types. Interestingly, extension of the phase of neural progenitor expansion during the evolution of brain is considered as major advantage for primates, as it added extra layers in the cortex. miRNAs are major determinants of the choice between symmetric versus asymmetric cell division.

The role of the miR-34/449 family of highly conserved miRNAs in asymmetric cell division was found in a screen designed to identify miRNAs that can affect the orientation of the mitotic spindle and, therefore, the exit from the cell cycle by neuronal progenitors (Fededa et al. 2016). Some members of the family have not undergone any sequence change from zebrafish to humans, suggesting that similar mechanisms may operate in contexts outside mammalian neural development. In zebrafish, miR-34 is deposited in the maternal ooplasm, and its knockdown results in a midbrain–hindbrain boundary defect (Soni et al. 2013). Across the vertebrates, miR-34 has conserved targets in the Notch and Dll messenger RNA (mRNA) that have important roles in the differentiation of the neural cell lineages. In mice, knockdown of miR-34 reduced the thickness of the cortex, an effect that is directly linked to the reduced number of cells in the cortical layers. Mimics of miR-449, a member of the miR-34 family, could knock down the cell adhesion protein JAM-A, resulting in disoriented spindles and cells that spend a longer time in mitosis (Figure 7.2) (Fededa et al. 2016).

The TRIM32 protein is another example of asymmetric distribution leading to daughter cells with specialized fates. The dividing cells in the ventricular zone sometimes give rise to daughter cells with high (or low) levels of TRIM32 protein that go on to differentiate into neurons. The TRIM32 protein binds to the Argonaute protein, Ago1, which in turn influences the levels of several miRNAs simultaneously.

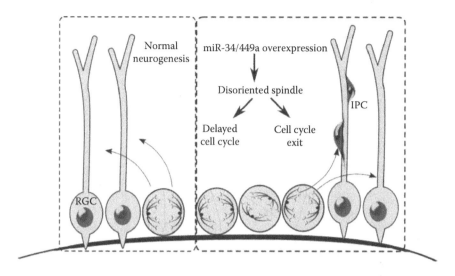

FIGURE 7.2 miRNAs control spindle orientation and cell cycle length. The schematic representation shows that during normal neurogenesis, RGCs divide with the orientation of the mitotic spindle horizontal relative to the ventricular surface, which leads to symmetric division and generates new RGCs. In the miR-34/449a overexpression condition, the mitotic spindle orientation changes to oblique or vertical to the plane of division, due to which the cells divide asymmetrically. IPC, intermediate progenitor cell.

Let-7, well known as a miRNA discovered early on in *Caenorhabditis elegans* and subsequently implicated in several cancers, is among the miRNAs that are bound to the TRIM32–Ago1 complex. Knockdown of Let-7 results in the accumulation of Nestin-positive neural stem cells, an effect that can be achieved even more effectively by TRIM32 knockdown. It appears that the TRIM32–Ago1 complex enhances miRNAs such as Let-7, encouraging cells along the differentiation pathway. Daughter cells that receive less of the TRIM32 protein, by contrast, stay in the proliferative zone, propagating the stem cell state. It is interesting to note that miR-449, like Let-7, is in the TRIM32–Ago1 complex along with some other miRNAs that influence neural cell fate decisions (Schwamborn, Berezikov, and Knoblich 2009).

7.1.2 NEURAL CELL FATE DETERMINATION

Once the neural progenitor cell exits the cell cycle, it progressively expresses genes that specify its fate as a neuronal cell or a glial cell. Highly complex, interconnected networks between transcription factors and miRNAs establish the fate of the cell, locking the cell into a certain expression profile by systematically switching on, shutting down, and tuning the expression of genes. The cascade of gene expression that specifies the alternative fates of the two main chemosensory neurons in *C. elegans* ASEL and ASER has become a model for understanding the mutually reinforcing roles of transcription factors and miRNAs (Johnston and Hobert 2003). Oliver Hobert's comparison of the features of transcription factors and miRNAs is useful for developing a deeper understanding of the division of labor between these two main effectors of gene regulation. Both transcription factors and miRNAs can affect multiple targets and establish a temporal order of expression by their varying affinities for targets. For instance, the upregulation of a single miRNA, say M1, in the cell reduces the expression the target gene, T1, drastically. Further increase in number of the same miRNA M1 may effectively shut down a second target of relatively lower affinity, T2, thus, establishing a temporal order of repression among the targets depending on their affinity for the miRNA.

A recurring network motif in miRNA–transcription factor networks is the feedback loop. The transcription factor, TLX, and miR-9 are connected by such a feedback loop. TLX binds to sites downstream to the miR-9 locus and represses the expression of the miRNA. miR-9 in turn binds to target sites in the TLX mRNA and inhibits its expression (Figure 7.3).

Ectopic overexpression of miR-9 resulted in fewer proliferating cells and increased migration of neurons in the cortical plate, showing that miR-9 accelerates neural differentiation and inhibits stem cell proliferation (Zhao et al. 2009).

An interesting theme in regulatory pathways that establish alternative neuronal and glial cell fates is the deployment of a miRNA to suppress a specific transcript isoform. For instance, miR-124 suppresses PTBP1, the glial isoform of the splicing factor, while PTBP2, the neural isoform with an altered specificity, drives alternative neuronal splicing. Ectopic expression of miR-124 can therefore drive the expression of the neuronal transcript variants of a large number of genes even in nonneuronal cells such as cultured HeLa cells (Makeyev et al. 2007). These neuronal fate-determining miRNAs are themselves under the regulation of transcription factors such as REST/coREST, CREB, and MeCP2. For instance, miR-124 is regulated by the REST/coREST complex (Conaco et al. 2006). Similarly, isoform-specific processing of Let-7 by Lin28 is important for neural stem cell commitment (Kawahara et al. 2011). Alternative 3′ untranslated (UTR) regions are, thus, the cell's way of modulating miRNA susceptibility, and miRNA family members may share the same seed, yet be processed differently to produce cell type-specific expression patterns.

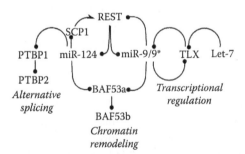

FIGURE 7.3 miRNAs and transcription factor form regulatory networks in neurogenesis. The REST complex inhibits expression of miR-9 and miR-124. The transcription factor TLX suppresses miR-9. TLX is also a downstream target of Let-7. The negative feedback loop by miR-9 suppresses both REST and TLX. miR-124 inhibits the splicing factor PTBP1, derepressing PTBP2, which leads to neuron-specific alternative splicing and differentiation.

7.2   MicroRNAs IN NEURAL CELL MIGRATION

Young neurons set out from their place of origin and travel through a spatial maze marked by chemical signals in the form of growth factors, developmental morphogens, and hormones. Much like a child starts out with latent abilities in a variety of areas but as it grows, experiences and external signals shape this landscape, limiting possibilities and committing it to a certain area of expertise, neurons also commit themselves to specific roles under the tutelage of signaling molecules, during the coincidental processes of maturation and migration.

The miRNA miR-128 plays a critical role in controlling the migration of cortical neurons. The precursor form of this miRNA, pre-miR-128, is expressed abundantly in the embryonic mouse, between day 12.5 (E12.5) to postnatal day 3. However, the processing of the precursor seems to be kept in check such that mature miR-128 appears only from E16.5 to coincide with the period of active neurogenesis. Spatially, the expression of the precursor, pre-miR-128, is more abundant in the deep layers of the cortex and in the adult neurogenic niches. In contrast, the mature miRNA, miR-128, is expressed in the postmigratory neurons found in the outer layers of the cortex. When a sponge was used to reduce the level of miR-128 in these cells, they migrated rapidly to the outer layer (Figure 7.4).

On the other hand, overexpression of the miRNA resulted in sluggish migration, with more neurons of aberrant morphology being found in the middle layers of the cortex. This phenotype was reminiscent of the defects seen in patients of the Börjeson–Forssmann–Lehmann syndrome who show a range of symptoms, prominent among them being mental retardation. Mutations in the zinc finger transcription factor, PHF6, are known to

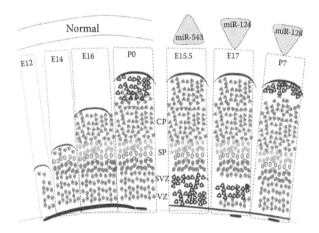

FIGURE 7.4 Dysregulation of miRNAs leads to defects in neuronal migration, integration, and cortical layer formation. Different layers of a normal mouse cortex during the development are shown in the left side with proliferating RGCs in the ventricular zone (VZ), intermediate progenitor cells in the subventricular zone (SVZ), migrating basal progenitors and immature neurons in the subplate (SP), and mature neurons in the cortical plate (CP). Overexpression of miR-543 prematurely starts neuronal differentiation, with significant decrease in RGCs in the VZ and an increase in neurons in the CP. miR-124 knockdown blocks the migration of neurons (shown as open triangles) from SP to CP. miR-128 knockdown shifts neurons (shown as open triangles) to the outermost layers of CP.

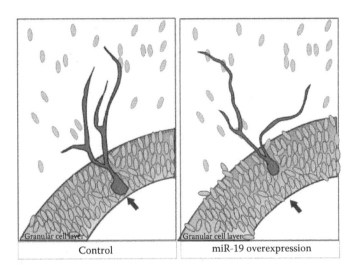

Control | miR-19 overexpression

FIGURE 7.5 miRNAs control the migration of newborn neurons. miR-19 overexpression changes the relative position of neurons within the GCL in the adult hippocampus as pointed by the arrows.

cause this disease. The Phf6 3′ UTR also carries target sites for miR-128, suggesting that direct targeting of Phf6 by miR-128 is a key step in regulation of neuronal migration. Overexpression of Phf6 effectively rescues the defect caused by overexpression of miR-128. Thus, the targeting of Phf6 by miR-128 and the proper timing of miR-128 processing is critical for the morphology and migration of neurons (Franzoni et al. 2015). Similarly, miRNAs from the miR-379–410 cluster control the expression of N-cadherin, a gene critical in the differentiation and migration of neurons (Rago et al. 2014). Here, too, the overexpression of the miRNAs promotes rapid migration, while knockdown of the miRNA results in impaired migration. Fred Gage, who pioneered the discovery of adult neurogenesis, has shown that the targeting of Rapgef2 by miR-19 is critical for the proper migration of hippocampal neurons. Neurons overexpressing miR-19 tend to reside in the inner layers of the granule cell layer (GCL) in the dentate gyrus (Figure 7.5), one of the sites of continued neurogenesis in the adult hippocampus (Han et al. 2016).

Dysregulation of miRNAs such as miR-9, miR-124, and miR-128 that have strong roles in neural cell differentiation (see previous section) also results in defects in neuronal morphology and migration (Delaloy et al. 2017; Franzoni et al. 2015; Volvert et al. 2017). All the miRNAs implicated in neuronal migration also show defects in neuronal morphology, typically affecting the number, density, and complexity of dendritic branches.

7.3 AXONOGENESIS, DENDRITOGENESIS, AND NEURITOGENESIS

Neurons are highly polarized cells with specific morphology linked to their functional role. The long, single axon; multiple short dendrites; and the tiny neurites that decorate their surface collectively give the neuron its distinctive shape. These regions of the neuron serve as subcellular compartments with local regulation that accounts for important neural phenomena such as learning and memory formation.

The dynamic nature of neuronal morphology makes it one of the aspects of neuronal function that can be tuned by modulators of synaptic activity.

The sciatic nerve injury model has been used extensively to study the role of various factors in promoting axonogenesis after injury. Using a conditional knockout of Dicer, it has been shown that following injury, regeneration was compromised due to the absence of miRNAs (Wu et al. 2012). A number of miRNAs are also known to be enriched in the axon, notably, the brain-enriched miR-181a-1* and miR-532. miR-9 not only is localized to the axons of cortical neurons but also regulates the expression of Map1b (Dajas-Bailador et al. 2012; Sasaki et al. 2014). Map1b is also regulated by miR-181d, which is delivered to the growth cone by FMRP, the fragile-X mental retardation protein (Wang et al. 2017). Both miRNAs, miR-9 and miR-181d, reduce axon length, since inhibition of these miRNAs by using antisense oligonucleotides leads to 20–30% lengthening of the axon (Hébert et al. 2008; Wang et al. 2017). The miRNAs that promote the formation of longer axons also tend to restrict the number of branches, resulting in a longer distance traversed and less complexity.

The dendrites of a neuron make connections, commonly with axons and rarely with dendrites of other neurons. The dendrites often spread out, forming many branches and subbranches, much like a tree, by a process called dendritic arborization. The more arborization there is, the higher the capability of the neuron to connect to many other neurons. Several miRNAs have been shown to modulate dendritic arborization in *in vitro* studies by using primary neurons. One of the best-studied miRNAs that affect dendritic arborization and is known to be localized to dendrites is miR-132. Acute loss of miR-132 engineered through the use of stereotactic injection of cre-expressing retrovirus in the mouse brain resulted in reduced dendrite length, branching, and spine density (Magill et al. 2010).

Neurites include both axons and dendrites, especially in the undifferentiated stage. Dendrites are covered with bulb-like membranous projections, of distinct shapes such as stubby, mushroom, and thin. These structures segregate a region of the cytoplasm with the necessary factors for the local translation of mRNAs. Some miRNAs are found in the spine, while others alter the dendritic spine density on overexpression or downregulation. The activity of these miRNAs then leads to the stabilization or eviction of mRNAs that are critical to the strengthening of synapses and eventually modify the connectivity of the neuron. One of the earliest miRNAs reported to affect dendritic spine density is miR-134. Schratt et al. (2006) showed that miR-134 was localized to the dendritic spine. Neurons overexpressing miR-134 had thinner, smaller spines, with less volume (Fiore et al. 2014; Schratt et al. 2006). Using brain-derived neurotrophic factor (BDNF) to increase the spine density and volume, it has been shown that several mRNAs gets translationally modulated during dendritogenesis. Among these, the Lim-domain-containing protein kinase 1 (Limk1), which regulates actin filament dynamics, has multiple conserved miR-134 binding sites. miR-134 overexpression resulted in translational repression of the Limk1 3′ UTR, an effect that could not be seen when the target site was mutated. Thus, it was clear that miR-134 was involved in local translational regulation, resulting in smaller spines (Schratt et al. 2006). Similarly, miR-132 also modulates spine density by regulating other factors involved in actin dynamics (Magill et al. 2010). However, a striking difference between miR-132 and the other spine density-modulating miRNAs such as miR-134, miR-29a/b, miR-181, and miR-125b is that miR-132 alone increases synaptic strength, whereas the others exert a repressive effect. The relation between dendritic spine morphology and synaptic strength is neither direct nor linear. miR-138 is notably unique in its ability to enlarge the dendritic spine, but it reduces synaptic strength (Siegel et al. 2009). It is important to bear in mind that miRNAs are downstream, local modulators of spine density and size, but their activity is under the control of neurotransmitter signaling. Thus, activity-dependent spine structure dynamics is brought about by

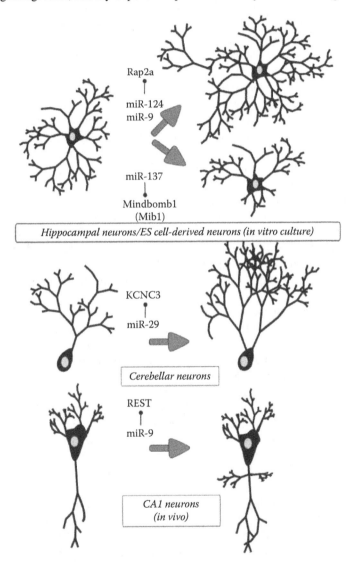

FIGURE 7.6 miRNAs control dendritic growth and arborization of immature neurons. miR-9 together with miR-124 increases dendritic arborization of embryonic stem (ES) cells derived neurons, while miR-137 reduces the dendritic complexity of maturing neurons. miR-9 independently controls dendritic development by repressing REST. miR-29 increases the dendritic arborization of cerebellar neurons.

the action of neurotransmitters and facilitated by the local miRNA-mediated control of translation of factors that in turn regulate actin dynamics (Figure 7.6).

But how are miRNAs modulated by neuronal activity? Neuronal excitation using a variety of pharmacological agents, for instance, bicuculline will induce the expression of miR-132 specifically. Neuronal stimulation also clears the dendrite off many miRNAs that show a rapid turnover rate following stimulation. It is thought that this activity-dependent clearing of a number of miRNAs, and the simultaneous transcriptional induction of selected miRNAs such as miR-132 and miR-134, is a mechanism for dynamically modulating the spine architecture.

So far, we have considered each stage of neurogenesis, starting with the proliferation of stem cells, to cell fate determination, migration, morphological changes, maturation, and finally synaptic activity-dependent changes. It is clear that some core neuronal miRNAs such as miR-124, miR-29a/b, miR-132, and miR-134 play roles in several stages of neurogenesis. Presumably, these miRNAs transition with the development of the neuron itself, taking up new targets as the cell progresses through the different stages of neurogenesis. The multivalent interactions between mRNAs that carry sites for multiple miRNAs and the partitioning of the miRNA pool between targets is likely to determine which targets are affected at any particular stage. While it is an impressive feat of molecular choreography, it makes therapeutic development all the more challenging. In the future, miRNA–target interaction networks that consider parameters of affinity and concentration may allow identification of hubs that can be targeted for optimal, programmable effects.

7.4 MicroRNAs AND NEURODEGENERATION

Neurodegenerative diseases form a large and growing fraction of incurable, debilitating, and chronic afflictions of the human population, especially with the rise in life expectancy. Mutations, even in ubiquitously expressed genes, can manifest as vulnerabilities in the brain. The most common are mutations that cause protein aggregation-related disorders wherein the cell struggles with clearing toxic clumps of proteins and eventually dies, which further triggers an exaggerated response from glial cells. The resulting inflammation and irreversible loss of cells manifests as cognitive decline, memory loss, and motor deficits. Besides being candidate biomarkers for early diagnosis, miRNAs have been proposed as potentially therapeutic agents that can be used to delay if not reverse the process of neurodegeneration (Table 7.2). A more ambitious goal is to harness our knowledge of miRNAs in embryonic neurogenesis to promote adult neurogenesis and replace the lost neurons.

Polyglutamine diseases are a class of protein aggregation-related neurodegenerative diseases caused by the expansion of CAG triplet repeats in the coding regions of genes, resulting in proteins with abnormally long polyglutamine domains. Interestingly, the expansion of CAG repeats in noncoding regions can also cause neurodegenerative diseases, implying that protein aggregation may be a consequence of and not the cause of neurodegeneration. Although the precise mechanism by which polyglutamine protein aggregates cause disease is not clear, the process of neurodegeneration is associated with the dysregulation of some miRNAs (Roshan et al. 2009). miR-29a/b is notable because its downregulation is a common factor in many neurodegenerative diseases, irrespective of the cause, be it CAG repeat expansion, other genetic mutations, or even

TABLE 7.2
Summary of miRNA–Target Interactions Implicated in Neuronal Dysfunction

Spinocerebellar Ataxia	Schizophrenia	Autism Spectrum Disorders	Epilepsy	Huntington's Disease	Parkinson's Disease	Alzheimer's Disease	Stroke	Cerebral Ischemia	miRNA/Disease
	SHANK3 (Choi et al. 2015)	SHANK3 (Choi et al. 2015)							Let7b
						Mmu, Hu TLR7 (Lehmann et al. 2012)			Let7f
							IGF-1 (Selvamani et al. 2012)		miR-1
							IGF-1 (Selvamani et al. 2012)		miR-7
				Hu REST/CoREST (Packer et al. 2008) CREB, BDNF (Müller 2014)		Hu BACE1 (Hébert et al. 2008)			miR-9/9*
									miR-10b-5p
								Rat Bcl2 (Shi et al. 2013)	miR-15b

(Continued)

TABLE 7.2 (CONTINUED)
Summary of MiRNA–Target Interactions Implicated in Neuronal Dysfunction

Spinocerebellar Ataxia	Schizophrenia	Autism Spectrum Disorders	Epilepsy	Huntington's Disease	Parkinson's Disease	Alzheimer's Disease	Stroke	Cerebral Ischemia	MiRNA/ Disease
ATXN1 (Lee et al. 2008)									miR-16
									miR-19
	Mmu VDAC (Roshan et al. 2014)			Mmu (Lee et al. 2011)		Mmu APP, ERK1, Tau (Liu et al. 2012)			miR-22
	(1) Mmu VDAC (Roshan et al. 2014); (2) cellular model BACE1 (Roshan et al. 2012) Hu (Perkins et al. 2007)					Hu BACE1 (Lei et al. 2015)		Rat DNMT3A (Pandi et al. 2013)	miR-29c
						Hu BACE1 (Hébert et al. 2008)			miR-29a/b
	Hu (Perkins et al. 2007)			CREB BDNF (Müller 2014)					miR-30a-5p miR-30e

(Continued)

TABLE 7.2 (CONTINUED)
Summary of MiRNA–Target Interactions Implicated in Neuronal Dysfunction

MiRNA/Disease	Spinocerebellar Ataxia	Schizophrenia	Autism Spectrum Disorders	Epilepsy	Huntington's Disease	Parkinson's Disease	Alzheimer's Disease	Stroke	Cerebral Ischemia
miR-34a		SHANK3 (Choi et al. 2015)	SHANK3 (Choi et al. 2015)						
miR-34b					Hu (Gaughwin et al. 2011)	Hu (Miñones-Moyano et al. 2011)			
miR-34c						Hu (Miñones-Moyano et al. 2011)	Mmu, Hu (Zovoilis et al. 2011)		
miR-98-5p							Cellular model SNX6 (Li et al. 2016)		
miR-101	Cellular model ATXN1 (Lee et al. 2008)								
miR-107							Hu BACE1 (Nelson and Wang 2010)		

(Continued)

TABLE 7.2 (CONTINUED)
Summary of MiRNA–Target Interactions Implicated in Neuronal Dysfunction

Spinocerebellar Ataxia	Schizophrenia	Autism Spectrum Disorders	Epilepsy	Huntington's Disease	Parkinson's Disease	Alzheimer's Disease	Stroke	Cerebral Ischemia	MiRNA/ Disease
Cellular model ATXN1 (Lee et al. 2008)	Hu, Mmu DNMT3A, GATA2, and DPYSL3 (Miller et al. 2012)					(1) Cellular model BACE1 (Fang et al. 2012); (2) *Drosophila* Alzheimer model delta (Kong et al. 2015b)			miR-124
					Mmu ERK2 regulatory loop (Tan et al. 2013)				miR-128
									miR-130
						ITPKB (Salta et al. 2016)		MeCP2 (Lusardi et al. 2009)	miR-132

(Continued)

TABLE 7.2 (CONTINUED)
Summary of MiRNA–Target Interactions Implicated in Neuronal Dysfunction

Spinocerebellar Ataxia	Schizophrenia	Autism Spectrum Disorders	Epilepsy	Huntington's Disease	Parkinson's Disease	Alzheimer's Disease	Stroke	Cerebral Ischemia	MiRNA/Disease
			Cell culture model Limk1 (Wang et al. 2014)				Rat Limk1 (Liu et al. 2017)		miR-134
					Drosophila α-synuclein (Kong et al. 2015a)				miR-137
							Rat superoxide dismutase-2 (Dharap et al. 2009)		miR-145
						Hu Mmu 2AG (Zhang et al. 2014)	Mmu GRP78 (Ouyang et al. 2012)		miR-181
									MiR-188-3p
								Rat SIRT1 (Xu et al. 2012)	miR-199a

(Continued)

TABLE 7.2 (CONTINUED)

Summary of MiRNA–Target Interactions Implicated in Neuronal Dysfunction

Spinocerebellar Ataxia	Schizophrenia	Autism Spectrum Disorders	Epilepsy	Huntington's Disease	Parkinson's Disease	Alzheimer's Disease	Stroke	Cerebral Ischemia	MiRNA/Disease
								PHD2/HIF-1 (Rink and Khanna 2011)	miR-200
			Hu ICAM1 (Kan et al. 2012)						miR-221
			Hu ICAM1 (Kan et al. 2012)						miR-222
	SHANK3 (Choi et al. 2015)	SHANK3 (Choi et al. 2015)					Mmu GluR2/NR2B (Harraz et al. 2012)		miR-223
									miR-504

Note: Each row pertains to a specific miRNA, the intersection of a row and column mentions the target gene, and the reference to the original report is in barckets. Hu, human sample; Mmu, mouse model used in the study.

Legend:

miRNA downregulated in disease

miRNA upregulated in disease

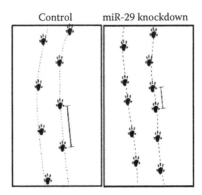

FIGURE 7.7   miRNA in neurodegenerative disease. Knockdown of miR-29 in mouse brain results in behavioral changes, here depicted by a pictorial representation of footprint analysis. The hind legs of the control or miR-29 knockdown mice were inked and allowed to walk on narrow paper. The distance between adjacent hind legs was measured.

viral infections (Roshan et al. 2012). As mentioned earlier, miR-29 family members are enriched in the axon and their knockdown is associated with neuronal apoptosis and defects in dendritic arborization. It can also exert long-term effects by affecting DNA methyltransferases, thus setting long-term epigenetic changes in motion. We have shown that transient, brain-specific knockdown of miR-29a/b was sufficient to cause the differential expression of target genes in the apoptotic pathway and a rapid loss of neurons in the mouse brain. Further, these mice displayed symptoms of ataxia, with disoriented, shorter steps (Figure 7.7) (Roshan et al. 2014).

The circulating form of the same miRNA is also reduced in expression in Alzheimer's patients. It is not clear if the ability of this miRNA, *in vitro*, to target the beta-site amyloid precursor protein cleaving enzyme 1 (BACE1) is indeed responsible for aggravating the pathology. Recently, it has been shown that recombinant pre-miR-29b can modulate human BACE1 in mouse neuroblastoma cells expressing the beta-amyloid peptide, projecting it as a novel biopharmaceutical product.

A bold attempt to use *in vitro*-harvested exosomes to deliver miR-124 to the brain of Huntington's disease model mice has shown that although modest increase in the miRNA could be achieved, known targets in neurogenesis were unaffected (Lee et al. 2017). It also failed to show any modification of behavioral symptoms. The replacement of dying neurons by promoting adult neurogenesis could be under multiple mechanisms of repression, and the expression of a single miRNA might not be sufficient to unlock these barriers. On the other hand, the ectopically delivered miRNA might have been quenched away by abundant, high-affinity targets or rendered inaccessible to the intended target due to intracellular compartmentalization.

7.5   MicroRNAs AS THERAPEUTICS AND BIOMARKERS

miRNAs play important roles as regulators of important protein-coding genes. They allow rapid response to environmental signals and through their own turnover can quickly release their control on mRNAs, effectively resetting conditions. However, many miRNAs individually bring about only a 10–40% change in the target gene expression. Although there are exceptions to this rule, it is generally accepted that miRNAs "tune" target gene expression rather than "switch (it) off." Their small size also makes it rarer to find mutations in the miRNA. Further, miRNAs can bind to targets with incomplete complementarity. Therefore, the involvement of miRNAs in human pathology is, generally, not through mutations that alter their expression, function, localization, or stability. A notable exception to this trend is the discovery of a single-nucleotide polymorphism (SNP) from a genome-wide association study on schizophrenia. This SNP maps to the intron of the primary transcript that gives rise to miR-137 (The Schizophrenia Psychiatric Genome-wide Association Study (GWAS) Consortium 2011). Since miR-137 is already known to regulate adult neurogenesis and neuronal maturation, it may provide some much-sought-after explanation of the etiology of the disease.

The chromosomal location 22q11.2 is known to harbor microdeletions linked to schizophrenia and other diseases with cognitive dysfunction. The deleted region includes several genes of importance to the neuronal system, including the miRNA biogenesis protein DGCR8 and the miRNA miR-185. The loss of DGCR8 in mouse models of the disease results in a loss of several miRNAs, and the loss of miR-185 is known to cause defects in dendritic spine formation (Sellier et al. 2014). Barring these examples, early reports of genomic rearrangements (such as those in Down syndrome) that result in the loss of specific neuronal miRNA have been disproven. Nevertheless, miRNA expression profiles give important information regarding the status of the cell. For instance, the expression of miRNAs can be biomarkers of cancer progression. Some miRNAs, by virtue of their presence in exosomes, can also provide plasma biomarkers, especially when the primary affected tissue is not accessible. As a generalization, it can also be argued that inhibition of miRNAs may have unintended consequences because of their differential effects on multiple targets. It is pertinent to note that this is equally true of an enzyme- or protein-interaction-based target. Offsetting this disadvantage is the small size of miRNAs, which renders them amenable to cellular delivery. Last but not the least, modified and stabilized miRNAs may be used to supplement downregulation due to mutations or aging. As transient regulators of gene expression, their levels can be externally tweaked and controlled, without causing irreversible changes. Small molecules that interact with RNA are well known, but only very few have been characterized extensively enough to be used as specific drugs in pathological conditions. Notably, ribocil is an early, if not the first, RNA-binding drug. It binds to the bacterial flavin mononucleotide riboswitch and represses the RibB mRNA, thus preventing bacterial pathogens from producing essential riboflavin (Howe et al. 2015). As a proof of concept of synthetic small-molecule-based discovery of druggable noncoding RNAs, this study reinforces the promise of RNA-structure-targeting therapeutics. There is ample scope for drug-repurposing studies using miRNAs as targets, and it is indeed expected that the empirical benefits and undesirable side effects of certain known drugs may indeed be attributed to their action on noncoding RNAs. Mimics of at least two miRNAs, miR-29 and miR-34, are in phase I clinical trials for scleroderma and solid tumors, respectively. Safety information from these studies may further accelerate the development of these miRNAs for neurodegenerative diseases.

7.6   PRIMATE-SPECIFIC MicroRNAs

The human brain has remarkably increased in size as well as acquired greater cognitive abilities compared with the brains of our ancestors, chimpanzees, or gorillas. It is the most complex organ of the body, specially evolved to perform higher-order tasks such as motor control, learning, memory, cognition, emotions, and behavior. Each of these tasks is performed by specialized brain areas and brain circuits. Any disruption in the wiring of brain circuits during or after brain development leads to neurodevelopmental or neuronal disorders. From the comparison between similarly sized or larger-sized brains, it was noticed that the human cortex has the maximum number of neurons (Herculano-Houzel 2012). Thus, it is not size of the human brain, but the number of neurons, their packaging, and the final architecture of the brain that might be the key to its cognitive abilities and complexities. How the human brain develops and integrates such a large number of neurons in a relatively small space, in a tightly regulated fashion, is the central question of neurobiology.

It is intuitively appealing to suggest that the unique features of the human brain maybe dependent on the evolution of primate-specific genes. Comparative genomics reveals the presence of primate-specific miRNAs, such as the C19MC cluster of 46 miRNAs that is expressed in the placenta, testis, and stem cells (Noguer-Dance et al. 2010). The imprinted cluster and its deregulation are implicated in pediatric glioblastoma owing to the ability of miRNAs to regulate cell proliferation. Even when miRNAs are conserved at the DNA level, several miRNAs adopt primate or even human-specific expression patterns in the rapidly evolving prefrontal cortex. These miRNAs may acquire new targets after getting edited or expressed in new regions of the brain. Some primate-specific miRNAs are expressed highly in the brain. miR-1202 presents an interesting case where the expression of this primate-specific miRNA is abundant in the human brain compared with monkey brains and enriched in the brain compared with other human tissues (Lopez et al. 2014). Following the standard approach of target prediction and validation, GRM4, a gene with a known function in synaptic transmission, was found to be a target of miR-1202. This miRNA, originally identified from postmortem brain tissue, was also reduced in the circulation of patients of major depressive disorder (MDD). Further, MDD patients who respond to drugs that modulate serotonergic response also showed restoration of the miRNA, suggesting that it is mechanistically linked to the disease and a biomarker for efficacy of treatment (Lopez et al. 2014).

In the past decade, miRNAs have greatly advanced our understanding of the brain function. At the same time, studies on neural cell types have also allowed us to appreciate the finer points of RNA biology. The fundamental processes by which miRNAs affect target genes are largely conserved across evolution. Some generalizations have indeed stood the test of time, spanning nearly two decades since the discovery of let-7. Each miRNA can regulate a clutch of target transcripts simultaneously, a strategy that allows coordinate regulation of genes. Each protein-coding transcript may carry multiple miRNA-binding sites, and transcript isoforms provide a platform for differential regulation by miRNAs. Besides alternative splicing and editing, neuronal transcripts are known to carry exceptionally long UTR regions, paving the way for increased complexity of gene regulation. The transcripts with multiple target sites are more severely regulated provided that they are spaced out to avoid steric hindrances. Yet other generalizations have been the subject of intense debate with evidence in support of alternative hypothesis. MiRNAs can affect the stability or translation of the target, but the factors that drive the choice between these alternative modes of posttranscriptional regulation remains elusive. Interestingly, neural cell types may favor different forms of miRNA-mediated regulation, perhaps due to the cell-type specific expression of RNA binding proteins in the brain. The division of labor between the four mammalian Argonaute proteins is unresolved. Lastly, we have yet to fully comprehend the role of miRNAs in long-term epigenetic reprogramming of the nervous system. The role of miRNAs in cognitive enhancement also remains underexplored. As we understand neuronal miRNAs in greater depth, we also uncover mechanisms by which their expression is regulated in response to environmental cues, metabolic signals from the gut, and physiological processes such as aging. In summary, the initial phase of the discovery of miRNAs and targets is now opening up new avenues for revealing novel, fascinating aspects of how our brain works.

REFERENCES

Berenguer, J., Herrera, A., Vuolo, L., Torroba, B., Llorens, F., Sumoy, L., and Pons, S. 2013. MicroRNA 22 regulates cell cycle length in cerebellar granular neuron precursors. *Molecular and Cellular Biology.* **33** (14): 2706–2717.

Bian, S., Hong, J., Li, Q., Schebelle, L., Pollock, A., Knauss, J. L., Garg, V., and Sun, T. 2017. MicroRNA cluster miR-17-92 regulates neural stem cell expansion and transition to intermediate progenitors in the developing mouse neocortex. *Cell Reports.* **3** (5): 1398–1406.

Choi, S.-Y., Pang, K., Kim, J. Y., Ryu, J. R., Kang, H., Liu, Z., Kim, W.-K., Sun, W., Kim, H., and Han, K. 2015. Post-transcriptional regulation of SHANK3 expression by microRNAs related to multiple neuropsychiatric disorders. *Molecular Brain.* **8** (1): 74.

Conaco, C., Otto, S., Han, J.-J., and Mandel, G. 2006. Reciprocal actions of REST and a microRNA promote neuronal identity. *Proceedings of the National Academy of Sciences of the United States of America.* **103** (7): 2422–7.

Dajas-Bailador, F., Bonev, B., Garcez, P., Stanley, P., Guillemot, F., and Papalopulu, N. 2012. MicroRNA-9 regulates axon extension and branching by targeting Map1b in mouse cortical neurons. *Nature Neuroscience.* **15** (5): 697–699.

Delaloy, C., Liu, L., Lee, J.-A., Su, H., Shen, F., Yang, G.-Y., Young, W. L., Ivey, K. N., and Gao, F.-B. 2017. MicroRNA-9 coordinates proliferation and migration of human embryonic stem cell-derived neural progenitors. *Cell Stem Cell.* **6** (4): 323–335.

Dharap, A., Bowen, K., Place, R., Li, L.-C., and Vemuganti, R. 2009. Transient focal ischemia induces extensive temporal changes in rat cerebral microRNAome. *Journal of Cerebral Blood Flow & Metabolism.* **29** (4): 675–687.

Edbauer, D., Neilson, J. R., Foster, K. A., Wang, C.-F., Seeburg, D. P., Batterton, M. N., Tada, T., Dolan, B. M., Sharp, P. A., and Sheng, M. 2010. Regulation of synaptic structure and function by FMRP-associated microRNAs miR-125b and miR-132. *Neuron.* **65** (3): 373–384.

Fang, M., Wang, J., Zhang, X., Geng, Y., Hu, Z., Rudd, J. A., Ling, S., Chen, W., and Han, S. 2012. The miR-124 regulates the expression of BACE1/β-secretase correlated with cell death in Alzheimer's disease. *Toxicology Letters.* **209** (1): 94–105.

Fededa, J. P., Esk, C., Mierzwa, B., Stanyte, R., Yuan, S., Zheng, H., Ebnet, K., Yan, W., Knoblich, J. A., and Gerlich, D. W. 2016. MicroRNA-34/449 controls mitotic spindle orientation during mammalian cortex development. *The EMBO Journal.* **35** (22): 2386–2398.

Fineberg, S. K., Kosik, K. S., and Davidson, B. L. 2009. MicroRNAs potentiate neural development. *Neuron.* **64** (3): 303–309.

Fiore, R., Rajman, M., Schwale, C., Bicker, S., Antoniou, A., Bruehl, C., Draguhn, A., and Schratt, G. 2014. miR-134-dependent regulation of Pumilio-2 is necessary for homeostatic synaptic depression. *The EMBO Journal.* **33** (19): 2135–2276.

Franzoni, E., Booker, S. A., Parthasarathy, S., Rehfeld, F., Grosser, S., Srivatsa, S., Fuchs, H. R., Tarabykin, V., Vida, I., and Wulczyn, F. G. 2015. miR-128 regulates neuronal migration, outgrowth and intrinsic excitability via the intellectual disability gene Phf6. *eLife.* **4**: e04263.

Gaughwin, P. M., Ciesla, M., Lahiri, N., Tabrizi, S. J., Brundin, P., and Björkqvist, M. 2011. Hsa-miR-34b is a plasma-stable microRNA that is elevated in pre-manifest Huntington's disease. *Human Molecular Genetics.* **20** (11): 2225–2237.

Giusti, S. A., Vogl, A. M., Brockmann, M. M., Vercelli, C. A., Rein, M. L., Trümbach, D., Wurst, W. et al. 2014. MicroRNA-9 controls dendritic development by targeting REST. *eLife.* **3**: e02755.

Gotz, M., and Huttner, W. B. 2005. The cell biology of neurogenesis. *Nature Reviews Molecular Cell Biology.* **6** (10): 777–788.

Han, J., Kim, H. J., Schafer, S. T., Paquola, A., Clemenson, G. D., Toda, T., Oh, J. et al. 2016. Functional implications of miR-19 in the migration of newborn neurons in the adult brain. *Neuron.* **91** (1): 79–89.

Harraz, M. M., Eacker, S. M., Wang, X., Dawson, T. M., and Dawson, V. L. 2012. MicroRNA-223 is neuroprotective by targeting glutamate receptors. *Proceedings of the National Academy of Sciences.* **109** (46): 18962–18967.

Hébert, S. S., Horré, K., Nicolaï, L., Papadopoulou, A. S., Mandemakers, W., Silahtaroglu, A. N., Kauppinen, S., Delacourte, A., and De Strooper, B. 2008. Loss of microRNA cluster miR-29a/b-1 in sporadic Alzheimer's disease correlates with increased BACE1/ beta-secretase expression. *Proceedings of the National Academy of Sciences of the United States of America.* **105** (17): 6415–6420.

Herculano-Houzel, S. 2012. The remarkable, yet not extraordinary, human brain as a scaled-up primate brain and its associated cost. *Proceedings of the National Academy of Sciences.* **109** (1): 10661–10668.

Howe, J. A., Wang, H., Fischmann, T. O., Balibar, C. J., Xiao, L., Galgoci, A. M., Malinverni, J. C. et al. 2015. Selective small-molecule inhibition of an RNA structural element. *Nature.* **526** (7575): 672–677.

Iyengar, B. R., Choudhary, A., Sarangdhar, M. A., Venkatesh, K. V., Gadgil, C. J., and Pillai, B. 2014. Non-coding RNA interact to regulate neuronal development and function. *Frontiers in Cellular Neuroscience.* **8**: 47.

Johnston, R. J., and Hobert, O. 2003. A microRNA controlling left/right neuronal asymmetry in Caenorhabditis elegans. *Nature.* **426** (6968): 845–849.

Kan, A. A., van Erp, S., Derijck, A. A. H. A., de Wit, M., Hessel, E. V. S., O'Duibhir, E., de Jager, W. et al. 2012. Genome-wide microRNA profiling of human temporal lobe epilepsy identifies modulators of the immune response. *Cellular and Molecular Life Sciences.* **69** (18): 3127–3145.

Kawahara, H., Okada, Y., Imai, T., Iwanami, A., Mischel, P. S., and Okano, H. 2011. Musashi1 cooperates in abnormal cell lineage protein 28 (Lin28)-mediated Let-7 family microRNA biogenesis in early neural differentiation. *The Journal of Biological Chemistry.* **286** (18): 16121–16130.

Kole, A. J., Swahari, V., Hammond, S. M., and Deshmukh, M. 2011. miR-29b is activated during neuronal maturation and targets BH3-only genes to restrict apoptosis. *Genes & Development.* **25** (2): 125–130.

Kong, Y., Liang, X., Liu, L., Zhang, D., Wan, C., Gan, Z., and Yuan, L. 2015. High throughput sequencing identifies microRNAs mediating α-synuclein toxicity by targeting neuroactive-ligand receptor interaction pathway in early stage of Drosophila Parkinson's disease model. *PLoS One.* **10** (9): e0137432.

Kong, Y., Wu, J., Zhang, D., Wan C., and Yuan, L. 2015. The role of miR-124 in Drosophila Alzheimer's disease model by targeting delta in notch signaling pathway. *Current Molecular Medicine.* **15** (10): 980–989.

Kosik, K. S. 2006. The neuronal microRNA system. *Nature Reviews Neuroscience.* **7** (12): 911–920.

Le, M. T. N., Xie, H., Zhou, B., Chia, P. H., Rizk, P., Um, M., Udolph, G., Yang, H., Lim, B., and Lodish, H. F. 2009. MicroRNA-125b promotes neuronal differentiation in human cells by repressing multiple targets. *Molecular and Cellular Biology.* **29** (19): 5290–5305.

Lee, S.-T., Chu, K., Im, W.-S., Yoon, H.-J., Im, J.-Y., Park, J.-E., Park, K.-H. et al. 2011. Altered microRNA regulation in Huntington's disease models. *Experimental Neurology.* **227** (1): 172–179.

Lee, S.-T., Im, W., Ban, J.-J., Lee, M., Jung, K.-H., Lee, S. K., Chu, K., and Kim, M. 2017. Exosome-based delivery of miR-124 in a Huntington's disease model. *JMD.* **10** (1): 45–52.

Lee, Y., Samaco, R. C., Gatchel, J. R., Thaller, C., Orr, H. T., and Zoghbi, H. Y. 2008. miR-19, miR-101 and miR-130 co-regulate ATXN1 levels to potentially modulate SCA1 pathogenesis. *Nature Neuroscience.* **11** (10): 1137–1139.

Lehmann, S. M., Kruger, C., Park, B., Derkow, K., Rosenberger, K., Baumgart, J., Trimbuch, T. et al. 2012. An unconventional role for miRNA: let-7 activates Toll-like receptor 7 and causes neurodegeneration. *Nature Neuroscience.* **15** (6): 827–835.

Lei, X., Lei, L., Zhang, Z., Zhang, Z., and Cheng, Y. 2015. Downregulated miR-29c correlates with increased BACE1 expression in sporadic Alzheimer's disease. *International Journal of Clinical and Experimental Pathology.* **8** (2): 1565–1574.

Li, Q., Li, X., Wang, L., Zhang, Y., and Chen, L. 2016. miR-98-5p acts as a target for Alzheimer's disease by regulating Aβ production through modulating SNX6 expression. *Journal of Molecular Neuroscience.* **60** (4): 413–420.

Liu, C., Teng, Z.-Q., Santistevan, N. J., Szulwach, K. E., Guo, W., Jin, P., and Zhao, X. 2010. Epigenetic regulation of miR-184 by MBD1 governs neural stem cell proliferation and differentiation. *Cell Stem Cell.* **6** (5): 433–444.

Liu, W., Liu, C., Zhu, J., Shu, P., Yin, B., Gong, Y., Qiang, B., Yuan, J., and Peng, X. 2012. MicroRNA-16 targets amyloid precursor protein to potentially modulate Alzheimer's-associated pathogenesis in SAMP8 mice. *Neurobiology of Aging.* **33**, (3): 522–534.

Liu, W., Wu, J., Huang, J., Zhuo, P., Lin, Y., Wang, L., Lin, R., Chen, L., and Tao, J. 2017. Electroacupuncture regulates hippocampal synaptic plasticity via miR-134-mediated LIMK1 function in rats with ischemic stroke. *Neural Plasticity.* **2017**: 9545646.

Lopez, J. P., Lim, R., Cruceanu, C., Crapper, L., Fasano, C., Labonte, B., Maussion, G. et al. 2014. miR-1202 is a primate-specific and brain-enriched microRNA involved in major depression and antidepressant treatment. *Nature Medicine.* **20** (7): 764–768.

Lusardi, T. A., Farr, C. D., Faulkner, C. L., Pignataro, G., Yang, T., Lan, J., Simon, R. P., and Saugstad, J. A. 2009. Ischemic preconditioning regulates expression of microRNAs and a predicted target, MeCP2, in mouse cortex. *Journal of Cerebral Blood Flow & Metabolism.* **30** (4): 744–756.

Lv, X., Jiang, H., Liu, Y., Lei, X., and Jiao, J. 2014. MicroRNA-15b promotes neurogenesis and inhibits neural progenitor proliferation by directly repressing TET3 during early neocortical development. *EMBO Reports.* **15** (12): 1305–1314.

Magill, S. T., Cambronne, X. A., Luikart, B. W., Lioy, D. T., Leighton, B. H., Westbrook, G. L., Mandel, G., and Goodman, R. H. 2010. MicroRNA-132 regulates dendritic growth and arborization of newborn neurons in the adult hippocampus. *Proceedings of the National Academy of Sciences.* **107** (47): 20382–20387.

Makeyev, E. V, Zhang, J., Carrasco, M. A., and Maniatis, T. 2007. The microRNA miR-124 promotes neuronal differentiation by triggering brain-specific alternative pre-mRNA splicing. *Molecular Cell.* **27** (3): 435–448.

Martynoga, B., Drechsel, D., and Guillemot, F. 2012. Molecular control of neurogenesis: A view from the mammalian cerebral cortex. *Cold Spring Harbor Perspectives in Biology.* **4** (10): a008359.

McNeill, E., and Van Vactor, D. 2012. MicroRNAs shape the neuronal landscape. *Neuron.* **75** (3): 363–379.

Miller, B. H., Zeier, Z., Xi, L., Lanz, T. A., Deng, S., Strathmann, J., Willoughby, D. et al. 2012. MicroRNA-132 dysregulation in schizophrenia has implications for both neurodevelopment and adult brain function. *Proceedings of the National Academy of Sciences.* **109** (8): 3125–3130.

Miñones-Moyano, E., Porta, S., Escaramís, G., Rabionet, R., Iraola, S., Kagerbauer, B., Espinosa-Parrilla, Y., Ferrer, I., Estivill, X., and Martí, E. 2011. MicroRNA profiling of Parkinson's disease brains identifies early downregulation of miR-34b/c which modulate mitochondrial function. *Human Molecular Genetics.* **20** (15): 3067–3078.

Mollinari, C., Racaniello, M., Berry, A., Pieri, M., de Stefano, M. C., Cardinale, A., Zona, C., Cirulli, F., Garaci, E., and Merlo, D. 2015. miR-34a regulates cell proliferation, morphology and function of newborn neurons resulting in improved behavioural outcomes. *Cell Death & Disease.* **6**: e1622.

Morgado, A. L., Xavier, J. M., Dionísio, P. A., Ribeiro, M. F. C., Dias, R. B., Sebastião, A. M., Solá, S., and Rodrigues, C. M. P. 2015. MicroRNA-34a modulates neural stem cell differentiation by regulating expression of synaptic and autophagic proteins. *Molecular Neurobiology.* **51** (3): 1168–1183.

Müller, S. 2014. In silico analysis of regulatory networks underlines the role of miR-10b-5p and its target BDNF in huntington's disease. *Translational Neurodegeneration.* **3** (1): 17.

Nelson, P. T., and Wang, W.-X. 2010. miR-107 is reduced in Alzheimer's disease brain neocortex: Validation study. *Journal of Alzheimer's Disease.* **21** (1): 75–79.

Neo, W. H., Yap, K., Lee, S. H., Looi, L. S., Khandelia, P., Neo, S. X., Makeyev, E. V., and Su, I.-h. 2014. MicroRNA miR-124 controls the choice between neuronal and astrocyte differentiation by fine-tuning Ezh2 expression. *Journal of Biological Chemistry.* **289** (30): 20788–20801.

Noguer-Dance, M., Abu-Amero, S., Al-Khtib, M., Lefèvre, A., Coullin, P., Moore, G. E., and Cavaillé 2010. The primate-specific microRNA gene cluster (C19MC) is imprinted in the placenta. *Human Molecular Genetics.* **19** (18): 3566–3582.

Ouyang, Y.-B., Lu, Y., Yue, S., Xu, L.-J., Xiong, X.-X., White, R. E., Sun, X., and Giffard, R. G. 2012. miR-181 regulates GRP78 and influences outcome from cerebral ischemia in vitro and in vivo. *Neurobiology of Disease.* **45** (1): 555–563.

Packer, A. N., Xing, Y., Harper, S. Q., Jones, L., and Davidson, B. L. 2008. The bifunctional microRNA miR-9/miR-9* regulates REST and CoREST and is downregulated in Huntington's disease. *The Journal of Neuroscience.* **28** (53): 14341–14346.

Pandi, G., Nakka, V. P., Dharap, A., Roopra, A., and Vemuganti, R. 2013. MicroRNA miR-29c down-regulation leading to de-repression of its target DNA methyltransferase 3a promotes ischemic brain damage. *PLoS One.* **8** (3): e58039.

Perkins, D. O., Jeffries, C. D., Jarskog, L. F., Thomson, M., Woods, K., Newman, M. A., Parker, J. S., Jin, J., and Hammond, S. M. 2007. microRNA expression in the prefrontal cortex of individuals with schizophrenia and schizoaffective disorder. *Genome Biology.* **8** (2): R27.

Pollock, A., Bian, S., Zhang, C., Chen, Z., and Sun, T. 2014. Growth of the developing cerebral cortex is controlled by microRNA miR-7 through modifying the p53 pathway. *Cell Reports*. **7** (4): 1184–1196.

Rago, L., Beattie, R., Taylor, V., and Winter, J. 2014. miR379–410 cluster miRNAs regulate neurogenesis and neuronal migration by fine-tuning N-cadherin. *The EMBO Journal*. **33** (8): 906–920.

Rink, C., and Khanna, S. 2011. MicroRNA in ischemic stroke etiology and pathology. *Physiological Genomics*. **43** (10): 521–528.

Roshan, R., Ghosh, T., Scaria, V., and Pillai, B. 2009. MicroRNAs: Novel therapeutic targets in neurodegenerative diseases. *Drug Discovery Today*. **14** (23–24): 1123–1129.

Roshan, R., Ghosh, T., Gadgil, M., and Pillai, B. 2012. Regulation of BACE1 by miR-29a/b in a cellular model of spinocerebellar ataxia 17. *RNA Biology*. **9** (6): 891–899.

Roshan, R., Shridhar, S., Sarangdhar, M. A., Banik, A., Chawla, M., Garg, M., Singh, V. P. A. L., and Pillai, B. 2014. Brain-specific knockdown of miR-29 results in neuronal cell death and ataxia in mice. *RNA*. **20** (8): 1287–1297.

Salta, E., Sierksma, A., Vanden Eynden, E., and De Strooper, B. 2016. miR-132 loss de-represses ITPKB and aggravates amyloid and TAU pathology in Alzheimer's brain. *EMBO Molecular Medicine*. **8** (9): 1005–1018.

Santos, M. C. T., Tegge, A. N., Correa, B. R., Mahesula, S., Kohnke, L. Q., Qiao, M., Ferreira, M. A. R., Kokovay, E., and Penalva, L. O. F. 2016. miR-124, -128, and -137 orchestrate neural differentiation by acting on overlapping gene sets containing a highly connected transcription factor network. *Stem Cells*. **34** (1): 220–232.

Sasaki, Y., Gross, C., Xing, L., Goshima, Y., and Bassell, G. J. 2014. Identification of axon-enriched microRNAs localized to growth cones of cortical neurons. *Developmental Neurobiology*. **74** (3): 397–406.

The Schizophrenia Psychiatric Genome-wide Association Study (GWAS) Consortium. 2011. Genome-wide association study identifies five new schizophrenia loci. *Nature Genetics*. **43** (10): 969–976.

Schratt, G. M., Tuebing, F., Nigh, E. A., Kane, C. G., Sabatini, M. E., Kiebler, M., and Greenberg, M. E. 2006. A brain-specific microRNA regulates dendritic spine development. *Nature*. **439**: 283–289.

Schwamborn, J. C., Berezikov, E., and Knoblich, J. A. 2009. The TRIM-NHL Protein TRIM32 activates microRNAs and prevents self-renewal in mouse neural progenitors. *Cell*. **136** (5): 913–925.

Sellier, C., Hwang, V. J., Dandekar, R., Durbin-Johnson, B., Charlet-Berguerand, N., Ander, B. P., Sharp, F. R., Angkustsiri, K., Simon, T. J., and Tassone, F. 2014. Decreased DGCR8 expression and miRNA dysregulation in individuals with 22q11.2 deletion syndrome. *PLoS One*. **9** (8): e103884.

Selvamani, A., Sathyan, P., Miranda, R. C., and Sohrabji, F. 2012. An antagomir to microRNA Let7f promotes neuroprotection in an ischemic stroke model. *PLoS One*. **7** (2): e32662.

Shi, H., Sun, B., Zhang, J., Lu, S., Zhang, P., Wang, H., Yu, Q. et al. 2013. miR-15b suppression of Bcl-2 contributes to cerebral ischemic injury and is reversed by sevoflurane preconditioning. *CNS & Neurological Disorders—Drug Targets*. **12** (3): 381–391.

Shi, X., Yan, C., Liu, B., Yang, C., Nie, X., Wang, X., Zheng, J., Wang, Y., and Zhu, Y. 2015. miR-381 regulates neural stem cell proliferation and differentiation via regulating Hes1 expression. *PLoS One*. **10** (10): e0138973.

Shibata, M., Kurokawa, D., Nakao, H., Ohmura, T., and Aizawa, S. 2008. MicroRNA-9 modulates Cajal–Retzius cell differentiation by suppressing Foxg1 expression in mouse medial pallium. *The Journal of Neuroscience*. **28** (41): 10415–10421.

Shin, J., Shin, Y., Oh, S.-M., Yang, H., Yu, W.-J., Lee, J.-P., Huh, S.-O. et al. 2014. miR-29b controls fetal mouse neurogenesis by regulating ICAT-mediated Wnt/[beta]-catenin signaling. *Cell Death & Disease*. **5**: e1473.

Siegel, G., Obernosterer, G., Fiore, R., Oehmen, M., Bicker, S., Christensen, M., Khudayberdiev, S. et al. 2009. A functional screen implicates microRNA-138-dependent regulation of the depalmitoylation enzyme APT1 in dendritic spine morphogenesis. *Nature Cell Biology.* **11** (6): 705–716.

Smrt, R. D., Szulwach, K. E., Pfeiffer, R. L., Li, X., Guo, W., Pathania, M., Teng, Z.-Q. et al. 2010. MicroRNA miR-137 regulates neuronal maturation by targeting ubiquitin ligase mind bomb-1. *Stem Cells.* **28** (6): 1060–1070.

Soni, K., Choudhary, A., Patowary, A., Singh, A. R., Bhatia, S., Sivasubbu, S., Chandrasekaran, S., and Pillai, B. 2013. miR-34 is maternally inherited in Drosophila melanogaster and Danio rerio. *Nucleic Acids Research.* **41** (8): 4470–4480.

Tan, C. L., Plotkin, J. L., Venø, M. T., von Schimmelmann, M., Feinberg, P., Mann, S., Handler, A. et al. 2013. MicroRNA-128 governs neuronal excitability and motor behavior in mice. *Science.* **342** (6163): 1254–1258.

Visvanathan, J., Lee, S., Lee, B., Lee, J. W., and Lee, S.-K. 2007. The microRNA miR-124 antagonizes the anti-neural REST/SCP1 pathway during embryonic CNS development. *Genes & Development.* **21** (7): 744–749.

Volvert, M.-L., Prévot, P.-P., Close, P., Laguesse, S., Pirotte, S., Hemphill, J., Rogister, F. et al. 2017. MicroRNA targeting of CoREST controls polarization of migrating cortical neurons. *Cell Reports.* **7** (4): 1168–1183.

Wang, B., Pan, L., Wei, M., Wang, Q., Liu, W.-W., Wang, N., Jiang, X.-Y., Zhang, X., and Bao, L. 2017. FMRP-mediated axonal delivery of miR-181d regulates axon elongation by locally targeting. *Cell Reports.* **13** (12): 2794–2807.

Wang, X.-M., Jia, R.-H., Wei, D., Cui, W.-Y., and Jiang, W. 2014. miR-134 blockade prevents status epilepticus like-activity and is neuroprotective in cultured hippocampal neurons. *Neuroscience Letters.* **572**: 20–25.

Wu, D., Raafat, A., Pak, E., Clemens, S., and Murashov, A. K. 2012. Dicer-microRNA pathway is critical for peripheral nerve regeneration and functional recovery in vivo and regenerative axonogenesis in vitro. *Experimental Neurology.* **233** (1): 555–565.

Xu, W.-H., Yao, X.-Y., Yu, H.-J., Huang, J.-W., and Cui, L.-Y. 2012. Downregulation of miR-199a may play a role in 3-nitropropionic acid induced ischemic tolerance in rat brain. *Brain Research.* **1429**: 116–123.

Zhang, J., Hu, M., Teng, Z., Tang, Y.-P., and Chen, C. 2014. Synaptic and cognitive improvements by inhibition of 2-AG metabolism are through upregulation of microRNA-188-3p in a mouse model of Alzheimer's disease. *The Journal of Neuroscience.* **34** (45): 14919–14933.

Zhang, W., Kim, P. J., Chen, Z., Lokman, H., Qiu, L., Zhang, K., Rozen, S. G., Tan, E. K., Je, H. S., and Zeng, L. 2016. MiRNA-128 regulates the proliferation and neurogenesis of neural precursors by targeting PCM1 in the developing cortex. *eLife.* **5**: e11324.

Zhao, C., Sun, G., Li, S., Lang, M.-F., Yang, S., Li, W., and Shi, Y. 2010. MicroRNA let-7b regulates neural stem cell proliferation and differentiation by targeting nuclear receptor TLX signaling. *Proceedings of the National Academy of Sciences.* **107** (5): 1876–1881.

Zhao, C., Sun, G., Li, S., and Shi, Y. 2009. A feedback regulatory loop involving microRNA-9 and nuclear receptor TLX in neural stem cell fate determination. *Nature Structural & Molecular Biology.* **16** (4): 365–371.

Zhong, W., Feder, J. N., Jiang, M.-M., Jan, L. Y., and Jan, Y. N. 2017. Asymmetric localization of a mammalian numb homolog during mouse cortical neurogenesis. *Neuron.* **17** (1): 43–53.

Zovoilis, A., Agbemenyah, H. Y., Agis-Balboa, R. C., Stilling, R. M., Edbauer, D., Rao, P., Farinelli, L. et al. 2011. MicroRNA-34c is a novel target to treat dementias. *The EMBO Journal.* **30** (20): 4299–4308.

8 MicroRNAs in the Progression of Hepatocellular Carcinoma

Kishor Pant, Amit Kumar Mishra, and Senthil Kumar Venugopal

CONTENTS

Introduction

Primary liver cancer of epithelial cell origin, either from hepatocytes or intrahepatic bile duct cells, is a major contributor to cancer incidences and mortalities. It is the sixth most commonly occurring cancer in the world and the third largest cause for cancer associated mortality. Liver cancers of non-epithelial origin are rare (El-Serag 2002). The most common histologic type of primary liver cancer, hepatocellular carcinoma (HCC), is a malignant tumor arising from hepatocytes, the liver's paren-chymal cells. Cytological features of hepatocytes and architecture of hepatic cords and sinusoids can be identified in these tumors. HCC is a primary liver malignancy and occurs mainly in patients with chronic liver disease and cirrhosis. HCC initi-ates with local growth, intrahepatic spread, and metastases to the distant organs. HCC is the third most common cause of mortality after lung and stomach cancer in patients. HCC is reported in the both men and women and is often diagnosed in people over 50 years of age. Every year approximately 750,000 new cases of HCC are reported worldwide, and a majority of the patients either require a liver transplant or die (Fattovich et al. 2004). Regarding India, information about HCC rates has not been updated since the 1988 to 2012 time span, due to people predominantly living in villages where a registry program for HCC patients and appropriate diagnostic techniques are inadequate (Acharya 2014).

8.1   EPIDEMIOLOGY

The distribution of HCC varies according to geographic locations. Disease burden is highest in areas with endemic hepatitis B virus (HBV) infection (where HBsAg prevalence is 8% or more), such as in sub-Saharan Africa and Eastern Asia, with incidence rates of over 20 per 100,000 individuals (El-Serag 2002; Yuen et al. 2009). Mediterranean countries such as Italy, Spain, and Greece have intermediate inci-dence rates of 10–20 per 100,000 individuals, while North and South America have a relatively low incidence (< 5 per 100,000 individuals). The global age distribution of HCC cases is related to dominant viral hepatitis in the underlying population and the age at which it was acquired. In regions of high incidence, the most common cause is HBV transmitted at birth. The diagnosis of HCC occurs about a decade earlier as compared with North America and Europe where the most common etiology is hepatitis C virus (HCV) acquired later in life. HCC is more common in men than women as HBV, HCV, and alcohol consumption are more prevalent and possibly more carcinogenic in males. In 80% to 90% of the cases, HCC occurs in the setting of cirrhosis (Fattovich et al. 2004; El-Serag 2004).

During the past two decades the numbers of HCC cases has increased about 80% in the United States. This increase has been mostly noticed in men, mainly in African-Americans having higher rates of HCC as compared to Caucasian men. This could be explained by the emergence of HCV during these periods, even though the increase in migration from the HBV-endemic nations may also have played some role. Conversely, other western countries including Italy, the United Kingdom, Canada, Japan, and Australia, have also noted related rising incidence of HCC (Goh et al. 2015). The increase in these countries is mainly reported amongst immigrants from other parts of the world with high prevalence of HCC, such as Asia and sub-Saharan Africa (Gomaa et al. 2008; El-Serag 2004).

8.2 HCC ETIOLOGY

Development of HCC is a complex process that involves various risk factors and modifications in a number of molecular pathways, as well as genetic mutations, which ultimately lead to malignant transformation and HCC disease progression (Fattovich et al. 2004). The main mechanisms involved in progression of hepatocellular carcinogenesis, and how these carcinogenic processes differ in relation to the different etiologic factors, are discussed below in detail (Figure 8.1).

8.2.1 CHRONIC HBV INFECTION

The role of HBV in the development of HCC is well known. Among the overall cases of liver cancer worldwide, HBV is responsible for 54% of HCC cases in patients, with a majority of cases in Asia, Africa, and the western Pacific region (Parkin 2006).

HBV infection causes hepatocyte injury and chronic inflammation, with subsequent increases in hepatocyte proliferation, fibrosis or cirrhosis. Constant regeneration

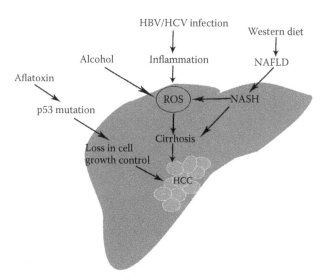

FIGURE 8.1   Etiological factors involved in the progression of hepatocellular carcinogenesis.

in cirrhosis may lead to increased turnover in hepatocytes and mutations in the genome that could result in genetic alterations, inactivation of tumor suppressor genes and activation of oncogenes in the host (Di Bisceglie 2009). HBV infection can also cause HCC in the absence of cirrhosis by integrating its DNA into host cells and acting as a mutagenic agent, causing chromosomal rearrangement and inducing genomic alteration. In addition, HBx is the HBV regulatory protein thought to activate genes involved in controlling cell proliferation, stimulating the protein kinase C and nuclear factor kappa B, deregulation of cell cycle control and inhibiting apoptosis (Cougot et al. 2005).

8.2.2 Chronic HCV Infection

HCV infected patients pose a 17-fold higher risk for developing HCC as compared to other risk factors (Hoshida et al. 2014). Of the liver cancer cases, 31.1% are attributed to the HCV infection, with higher prevalence in northern and middle Africa (Hoshida et al. 2014; de Oliveria Andrade et al. 2009). HCV infection causes chronic hepatic inflammation, cell proliferation, and cirrhosis in the host liver. HCV infection has also been reported to induce insulin resistance and led to the development of liver fibrosis and diabetes type 2 in patients (Gastaldi et al. 2017). The proinflammatory cytokine tumor necrosis factor (TNF) in the presence of HCV core protein was found to be responsible for the development of the insulin resistance. In addition, these proinflammatory cytokines are known to provoke Kupffer cell activation by the production of oxidative stress by HCV core protein exposure (Gastaldi et al. 2017).

8.2.3 Cirrhosis

Cirrhosis is the primary risk factor for HCC development; the major causes are HBV, HCV infection and alcohol abuse. Development of liver cirrhosis starts with long periods of chronic disease and is considered to reduce hepatocyte proliferation, representing the collapse in the regenerative capability of the liver (Fattovich et al. 2004). Cirrhosis is associated with an increase in fibrous tissue activation and a destruction of other liver cells, which are further responsible for the development of cancerous/tumorous nodules (Fattovich et al. 2004). Liver cirrhosis may have a significant impact on liver dysfunction and be responsible for the mortality rates associated with HCC.

8.2.4 Aflatoxin

The fungus of the genus, *Aspergillus* spp. is known to produce aflatoxin in mainly Asian and sub-Saharan Africa countries where temperature, climatic factors and food/grain storage methods permit fungus growth and lead to contamination in foods. Regions with high exposure of aflatoxin have reported a high prevalence of HCC. In HBV-infected patients it has also been noticed that intake of aflatoxin may increase the risk of development of HCC (Wu and Santella 2012). It has been reported that a high intake of aflatoxin and a high occurrence of HCC are present in areas with widespread HBV infection, and that patients who were exposed to both HBV and aflatoxin were at a higher risk of developing HCC (Kew 2013). In addition, aflatoxin is known to induce somatic mutations in the tumor suppressor p53 gene which is a common genetic abnormality in liver cancers and studies support a high level of mutation in p53 in HCC (Kew 2013).

8.2.5 Non-Alcoholic Fatty Liver Disease (NAFLD)

NAFLD is the most common liver ailment in developed countries with ~20% of individuals suffering. Even though, NAFLD occurs without abuse of alcohol, the hepatic histology shows akin to the alcoholic hepatitis and other changes in liver histology including steatosis (fat deposition), hepatic inflammation (Steatohepatitis), fibrosis and HCC (Baffy, Brunt, and Caldwell 2012). Progression of NAFLD is linked with metabolic syndromes, including obesity and type 2 diabetes (Margini and Dufour 2016) and this is considered to enhance the risk of developing many chronic liver diseases, cirrhosis, and HCC (Margini and Dufour 2016).

8.2.6 Alcohol Abuse

Alcohol drinking is the most common etiological factor of HCC and responsible for many mortalities worldwide (Kar 2014; Stroffolini 2005). In areas with low prevalence of HBV and HCV infection, alcohol consumption is a major risk factor of HCC. Alcohol may act as hepatocarcinogenic agent, also known to worsen liver damage by viral infection and support tumor growth (Mercer, Hennings, and Ronis 2015). Alcohol is a main cause of liver cirrhosis and may contribute from 15% to 45% of HCC cases in Western countries. It has a synergistic relationship with other risk factors such as hepatitis virus, diabetes, obesity, and smoking (Sidharthan and Kottilil 2014). According to some reports, as the consumption of alcoholic beverages keeps increasing worldwide, alcohol related liver disease and HCC is becoming a more serious issue for the scientific community (Testino et al. 2014).

8.3 miRNAs IN HCC

miRNA is small molecule (~22–25 nucleotides) that plays a significant role in the growth and development of many tumors in humans, including HCC. The expression of miRNAs regulates the development of HCC significantly and vice versa. Deregulation in the expression of the many miRNAs has been reported in the HCC and is briefly described below. Figure 8.2 and Table 8.1 illustrate the role of miRNAs in different stages of HCC development and progression.

8.3.1 Up-Regulated miRNAs in HCC

8.3.1.1 miRNA-21

miRNA-21, which is the most commonly upregulated miRNA in most cancerous tissues, is associated with the capacity of cancer cell migration and invasion in HCC and plays an important role in the regulation of anticancer drug sensitivity

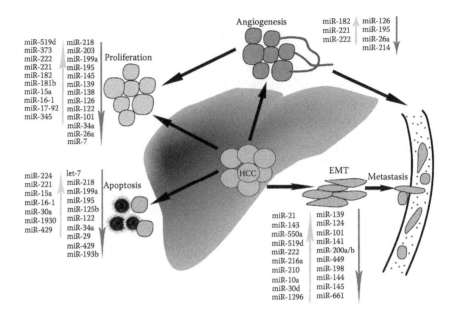

FIGURE 8.2 Role of miRNAs in different stages of HCC.

and resistance (Pan, Wang, and Wang 2010). 17q23, an encoding genetic locus, has been reported to be amplified in many solid tumors (Kasahara et al. 2002; Fujita et al. 2008). Steroid hormones, activator protein (AP-1), hypoxia and inflammation are some of the factors known to stimulate miR-21 expression (Fujita et al. 2008; Ribas and Lupold 2010). Key targets for miR-21 include tumor suppressors like phosphatases, programmed cell death-4 (PDCD4), and tensin homolog (PTEN) (Bao et al. 2013). MiR-21 mediated suppression of PTEN contributes to HCC cell proliferation, tumor growth and migration via epithelial–mesenchymaltransition (EMT) as well as activation of AKT/ERK pathways (Bao et al. 2013). In HCC, miR-21 and PDCD4 expression is inversely correlated (Zhu et al. 2012, 4).

8.3.1.2 miRNA-221 and miRNA-222

miRNA-221 and miRNA-222 are highly homologous in nature and have been shown to be upregulated in HCC. Cyclin dependent kinase (CDK) inhibitor p27, PTEN and tissue inhibitor of matrix metalloproteinase (TIMP)3 are the key targets for miRNA-221/222 (Yuan et al. 2013; Wong et al. 2010). miRNA-221/222 mediated suppression of PTEN and TIMP3 leads to activation of AKT in HCC, thus conferring metastatic potential to HCC cells (Garofalo et al. 2009). miRNA-221/222 targeted suppression of Cyclin dependent kinase (CDK) inhibitor p27 enhances cell proliferation, invasion and migration (Yuan et al. 2013).

TABLE 8.1
Deregulated miRNAs in HCC

miRNAs	Expression	Target Genes	Cellular Effects	References
miR-17-5p	↑	p38 pathway	Tumirogenesis and Invasion	(Ambros 2004)
miR-18a	↑	ERalpha	Increased cellular proliferation	(Liu et al. 2009)
miR-21	↑	PTEN, RECK, PDCD4,RHOB	Migration, Invasion, metastasis, Drug resistance	(Meng et al. 2007; Zhu et al. 2012:4; Connolly et al. 2010; Tommaru et al. 2010)
miR-135a	↑	FOXM1, MTSS1	Metastasis	(Liu et al. 2012)
miR-143	↑	FNDC3B	Metastasis	(Zhang et al. 2009)
miR-23a	↑	PGC-1α; G6PC	Gluconeogenesis	(Bo, Wang et al. 2012)
miR-155	↑	APC	Proliferation and Tumirogenesis	(Yiliang, Zhang et al. 2012)
miR-181b	↑	TIMP3	Proliferation, Tumirogenesis and Metastasis	(Bo, Wang et al. 2010)
miR-182	↑	MTSS1	Metastasis	(Jian, Wang et al. 2012)
miR-210	↑	VMP1; AIFM3	Proliferation, Metastasis, Apoptosis	(Ying et al. 2011; Wei, Yang et al. 2012)
miR-216a	↑	TSLC1	Tumirogenesis	(Chen et al. 2012)
miR-221	↑	CDKN1C/p57; DDIT4; Arnt	Apoptosis; Proliferation; Angiogenesis	(Santhekadur et al. 2012:1; Fornari et al. 2008a; Gramantieri et al. 2009)
miR-519d	↑	CDKN1A/p21; PTEN; AKT3; TIMP2	Proliferation; Invasion; Apoptosis	(Fornari et al. 2012)
MiR-550a	↑	CPEB4	Metastasis	(Tian et al. 2012)
miR-15a/16–1	↑	Bcl-2, cyclin D1, AKT3	Proliferation, Apoptosis	(Liu et al. 2013)

(Continued)

TABLE 8.1 (CONTINUED)
Deregulated miRNAs in HCC

miRNAs	Expression	Target Genes	Cellular Effects	References
miR-(17-92)	↑	c-Myc, E2F	Cell cycle apoptosis	(Aguda et al. 2008; Sylvestre et al. 2007)
miR-(106b-25)	↑	Bim, E2F1	Apoptosis, Proliferation	(Tan et al. 2014; Yang Li et al. 2009)
miR-1	→	ET-1	Proliferation and Metastasis	(Li et al. 2012)
miR-7	→	PIK3CD, mTOR, p/0S6K, CCNE1	Tumorigenesis, Metastasis, Cell cycle	(Xiao, Zhang et al. 2014:1; Fang et al. 2012)
miR-26a	→	CDK6, cyclin D1, PIK3C2α, c-MET	Metastasis and Angiogenesis	(Chai et al. 2013:2; Yang et al. 2013; Xin, Yang et al. 2014)
miR-29	→	Bcl-2, Bcl-w, Ras, matrix metalloproteinase-2 (MMP-2)	Apoptosis, Angiogenesis, Metastasis, Invasion	(Fang et al. 2011; Gebeshuber, Zatloukal, and Martinez 2009)
miR-34a	→	Cyclin D1, CDK4, and CDK2, c-Met	Proliferation, Apoptosis, Metastasis	(Na, Li et al. 2009; Cheng et al. 2010; Pengyuan, Yang et al. 2012:22)
miR-101	→	ZEB1, Rab5a, SOX9, FOS	EMT, Proliferation	(Yanqiong, Zhang et al. 2012:9; Shuai, Li et al. 2009; Sheng et al. 2014; Su et al. 2009)
miR-122	→	Bcl-w; Cyclin G1,ADAM17;Wnt1	Apoptosis; Metastasis; Angiogenesis	(Lin et al. 2008; Tsai et al. 2009; Coulouarn et al. 2009; Gramantieri et al. 2007)

(Continued)

TABLE 8.1 (CONTINUED)
Deregulated miRNAs in HCC

miRNAs	Expression	Target Genes	Cellular Effects	References
Let-7a	↓	Caspase-3	Apoptosis; Proliferation	(Zifeng, Wang et al. 2011; Tsang and Kwok 2008:7)
Let-7b	↓	HMGA2	Apoptosis; Proliferation	(Di Fazio et al. 2012:2)
Let-7c	↓	Bcl-xL; c-myc	Apoptosis; Proliferation	(Au et al. 2012; Zhu et al. 2011:7; Shah et al. 2007)
Let-7d	↓	—	Apoptosis; Proliferation	(Zifeng Wang et al. 2011)
Let-7f-1	↓	—	Apoptosis; Proliferation	(Zifeng Wang et al. 2011)
Let-7g	↓	Bcl-xL; COL1A2; c-Myc;p16(INK4A)	Apoptosis; Proliferation	(Shimizu et al. 2010; Lan et al. 2011)
miR-138	↓	CCND3	Cell cycle regulation	(Wen, Wang et al. 2012:3)
miR-139	↓	ROCK2; c-Fos	Proliferation, Metastasis	(Au et al. 2012; Wong et al. 2011; Fan et al. 2013)
miR-195	↓	CDK6, cyclin D1, CBX4, Wnt3a, VEGF, VAV2, CDC42	Apoptosis, Metastasis, Angiogenesis	(Ruizhi, Wang et al. 2013; Xu et al. 2009; Zheng et al. 2015:4; Yang, Yang et al. 2014:3)
miR-199a	↓	mTOR, PAK4, H1F1A, E2F3, DDR1	Cell growth, Apoptosis	(Murakami et al. 2006; Fornari et al. 2010; Shen et al. 2010)
miR-200family	↓	ZEB1, ZEB2, HNF-3β, Rho/ROCK, ASB4	EMT, Metastasis	(Dhayat et al. 2014)
miR-449	↓	c-Met, SIRT1	Metastasis	(Buurman et al. 2012; Hongyi, Zhang et al. 2014)
miR-520b	↓	MEKK2, cyclin D1	Cell cycle proliferation	(Weiying, Zhang et al. 2012:2)

8.3.1.3   miRNA-155

miRNA-155 plays an important role either as an oncomiRNA or as an oncosuppressor miRNA, depending on the tissue type in a wide variety of cancers. Key targets of miR-155 include RhoA and Ras homolog gene family members, and thus, facilitates growth factor (TGF) β-induced EMT and cell migration and invasion (Kong et al. 2008).

8.3.2   Downregulated miRNAs in HCC

8.3.2.1   miRNA-122

The liver-specific miR-122 plays an important role in regulating hepatocyte development, differentiation and metabolism. It is the most abundant miRNA in the liver and has been reported to be downregulated in HCC (Jopling 2012; J. Hu et al. 2012; Coulouarn et al. 2009; Lin et al. 2008). miRNA-122 and Cyclin G1 expression is inversely correlated in primary liver carcinomas indicating that cyclin G1 is a direct target of miR-122 (Gramantieri et al. 2007). Mouse models deficient in cyclin G1 were found to be less prone to developing hepatic tumors. Reports suggest that in HCC cells, miR-122/cyclin G1 interaction affects doxorubicin sensitivity by modulating p53 activity (Gramantieri et al. 2007). Efforts have been made to understand the effects of the association between miRNA-122 and the hepatitis virus, and recent advances indicate that miRNA-122 interaction with HCV RNA confers stability to it, as well as protects it against degradation (Li et al. 2014). While miRNA-122 is beneficial for HCV its effect on HBV is completely the opposite and downregulates HBV gene expression and replication. Chen et al. demonstrated that miR-122 targets HBV mRNA sequences coding for viral polymerase and the 3'-UTR region of the mRNA for the core protein of the HBV genome (Chen et al. 2011).

8.3.2.2   miRNA-101

miRNA-101 is a tumor suppressor miRNA, and is found to be significantly downregulated in multiple types of cancers, including HCC (Sheng et al. 2014). miRNA-101 targets myeloid cell leukemia sequence 1 (Mcl-1), a member of antiapoptotic B-cell lymphoma-2 (Bcl-2) family, and thus, displays its proapoptotic function (Su et al. 2009). In the hepatic cell line HepG2, miRNA-101 suppresses autophagy by directly targeting ATG4D, RAB5A, and STMN1 (Xu et al. 2013). Other downstream targets of miRNA-101 include ZEB1, DNMT3A, SOX9, v-fos FBJ murine osteosarcoma viral oncogene homolog (FOS), EZH2 and NLK (Li et al. 2009; Zhang et al. 2012, 9; Shen et al. 2014; Sheng et al. 2014). miRNA-101 mediated suppression of NLK, a MAP kinase related kinase, leads to reduced cancer cell growth and proliferation (Shen et al. 2014). miRNA-101-3p is downregulated by HBV, thus promoting HCC proliferation and migration (Sheng et al. 2014).

8.3.2.3   Let-7 Family

The let-7 miRNA family, an evolutionarily conserved miRNA family among many species, consists of 13 members and belongs to the family of tumor suppressor miRNAs (Reinhart et al. 2000, 7). It was first discovered in Caenorhabditis elegans

(Reinhart et al. 2000, 7). Let-7 miRNA family members, miRNA let-7c and let-7g have been shown to downregulate the expression of antiapoptotic protein Bcl-xL, a member of Bcl-2 family (Shimizu et al. 2010). HBx protein of HBV, suppresses let-7 family at a transcriptional level and upregulates expression of the signal transducer and activator of transcription-3, which in turn causes cell proliferation (Wang et al. 2010).

8.4 DEREGULATED miRNAs IN VIRAL HEPATITIS

The changing pattern of miRNAs during the HBV and HCC life cycles has been described in the Figure 8.3.

8.4.1 HBV

miRNAs miR-125a-5p, miR-199a-3p, and miR-210 interfere with the expression of HBsAg and reduce the amount of secreted HBsAg (Potenza et al. 2011; Zhang et al. 2010). miRNA-15a/miR-16-1, miRNA-17-92 cluster, and miR-224 target HBV mRNAs and were subsequently shown to be downregulated by HBV infection (Jung et al. 2013; Wang et al. 2013). HBV gene expression is promoted by miRNA-372/373, which has been shown to be upregulated in HBV infected liver tissues (Guo et al. 2011). In HCC, expression of miRNA-501 goes up and has been demonstrated to promote HBV replication by targeting HBXIP (Jin et al. 2013). In contrast miRNA-155 and miRNA-141 have been shown to inhibit HBV (Hu et al. 2012; Su et al. 2011).

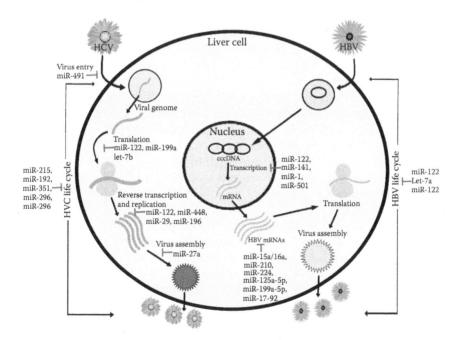

FIGURE 8.3   Changing patterns of miRNAs during HBV and HCC life cycle.

miRNA-155, an immune related miRNA, mediated activation of Janus kinase/ signal transducer and activator of transcription (JAK/STAT) signaling pathway leads to enhanced innate anti-viral immunity against HBV, and thus provides partial resistance against HBV infection (Su et al. 2011). On the other hand, ex vivo study in hepatoma cell-line HepG2 has revealed that miRNA-141 targets peroxisome proliferator-activated receptor alpha and suppresses HBV replication and transcription (Hu et al. 2012).

Novellino et al. demonstrated that miRNAs-27a, -30b, -122, -126, -145 and immunoregulatory miRNA-106b and miRNA-223 were present in the blood of HBV carriers along with circulating HBsAg and contributed towards HBV persistence as well as HCC (Novellino et al. 2012). Another study by Hayes et al. revealed that expression of miRNAs-122, -22 and -99a goes up significantly in the HBV infected patient sample (Hayes et al. 2012). The smallest HBV protein HBx, which is required for viral replication, but is not essential, has been reported to interplay between cellular miRNAs upregulation and downregulation (Tian et al. 2013). Table 8.2 demonstrates several miRNAs and their role in the progression HBV induced HCC.

8.4.2 HCV

A number of miRNAs have been identified to play a main role in regulating the virus replication and pathogenesis during the HCV life cycle. miRNA-122 is a liver-specific miRNA known to regulate HCV replication via inhibiting the HCV-RNA accumulation in infected cells. miRNA-122 binds directly to 5′ UTR of HCV-RNA and leads to inhibition of the 5′ terminal sequences from nucleolytic degradation, promoting HCV-RNA stability and proliferation in HCV genome (Jangra et al. 2010; Thibault et al. 2015; Conrad et al. 2013). Conversely, some other miRNAs including, miR-196, miR-199a, miR-448, miR-181c and let-7b have been reported to interact directly with HCV-RNA and inhibit HCV replication (Pedersen et al. 2007; Shrivastava et al. 2013).

HCV replication was also inhibited by the miR-448 (by targeting the NS5A coding region) and miRNA-196 (targeting the core region) via inhibiting the HCV genome. Overexpression of miR-199a inhibited HCV replication via binding to 5′UTR of thhe HCV genome at stem-loop II region (Murakami et al. 2006). In addition, Let-7b (Cheng et al. 2012) and miRNA-181c (Mukherjee et al. 2014) have also been reported to possess anti-HCV properties via directly targeting HCV genome. Table 8.3 and Figure 8.3 illustrate the regulation of miRNAs in HCV medicated pathogenesis of HCC.

The level of miRNA-29 was found to be downregulated in the liver of patients infected with HCV; however, overexpression of miR-29 inhibited viral HCV-RNA in HCV in hepatocytes (Bandyopadhyay et al. 2011). Inhibition of miR-29 has also been linked to the activation of HSCs and collagen synthesis (Bandyopadhyay et al. 2011). In HCV patients, the expression of miR-449a and miRNA-107 was observed to be decreased, but in alcoholic and non-alcoholic liver disease patients the miRNA level was unchanged (Sarma et al. 2012). These studies suggest that miRNAs can target genes involved in inflammation and fibrosis in patients with chronic HCV infection.

TABLE 8.2

Deregulated miRNAs in HBV-Induced HCC and Their Cellular Targets

miRNAs	Expression	Cellular Targets	Cellular Effects	References
miRNA-1	↓	HDAC4, MET	Proliferation (−)	(Datta et al. 2008; Zhang et al. 2011)
miRNA-15a	↓	BCL-2	Proliferation (−), Apoptosis (+)	(Liu et al. 2013; Yanling, Wang et al. 2013; Wu et al. 2011)
miRNA-16	↓	Cyclin D1, NCOR2	Proliferation (−), Apoptosis (+)	(Wu et al. 2011; Zeng et al. 2012)
miRNA-22	↓	CDKN1A, ERα, HDAC4	Proliferation (−)	(Jiang et al. 2011; Zhang et al. 2010; Shi and Xu 2013)
miRNA-23b	↓	MET	Proliferation (−), Migration (−)	(Salvi et al. 2009)
miRNA-26a	↓	IL-6, Cyclin D2, Cyclin E2	Proliferation (−), Metastasis (−)	(Yang et al. 2013; Chen et al. 2011)
miRNA-29c	↓	TNFAIP3	Proliferation (−), Apoptosis (+)	(Chun-Mei, Wang et al. 2011:3)
Let-7	↓	collagen type I a2, c-myc, Bcl-xL, RAS, HMGA2, NGF, STAT3	Proliferation (−), Migration (−), Apoptosis (+)	(Zeng et al. 2012; Yu, Wang et al. 2010; Johnson et al. 2005; Ji et al. 2010)
miRNA-34a	↓	CCL22, MET	Metastasis (−)	(Na, Li et al. 2009; Pengyuan, Yang et al. 2012:22)
miRNA-99a	↓	mTOR, IGF-1R	Proliferation (−)	(Li et al. 2011)
miRNA-199a-3p	↓	mTOR, PAK4	Proliferation (−), Apoptosis (+)	(Hou et al. 2011; Fornari et al. 2010)
miRNA-152	↓	DNMT1	Apoptosis (+), Metastasis (−), Aberrant DNA methylation	(Huang et al. 2010:1)
miRNA-148a	↓	MET, HPIP	Proliferation (−), Metastasis (−)	(J.-P. Zhang et al. 2014; Han et al. 2013; Xu et al. 2013)
miRNA-145	↓	HDAC2, ADAM17	Proliferation (−), Invasion(−)	(Xue-Wei, Yang et al. 2014; Noh et al. 2013:2)

(Continued)

TABLE 8.2 (CONTINUED)

Deregulated miRNAs in HBV-Induced HCC and Their Cellular Targets

miRNAs	Expression	Cellular Targets	Cellular Effects	References
miRNA-122	↓	ADAM10, Cyclin G1, Igf1R ADAM17, BCL-W	Proliferation (−), Apoptosis (+), Invasion (−)	(Gramantieri et al. 2007:1; Saifeng, Wang et al. 2012; Li et al. 2013:1; Coulouarn et al. 2009; Tsai et al. 2009)
miRNA-101	↓	DNMT3A, FOS, MCL-1, EZH2-	Metastasis (−), Apoptosis (+), Aberrant DNA methylation	(Su et al. 2009; Shuai, Li et al. 2009; Wang et al. 2014; Wei et al. 2013; Au et al. 2012)
miRNA-224	↑	API-5, Smad4	Proliferation (+), Metastasis (+)	(Lan et al. 2014; Scisciani et al. 2012; Wang et al. 2008)
miRNA-222	↑	PPP2R2A	Metastasis (+)	(Wong et al. 2010)
miRNA-155	↑	SOX6	Proliferation (+)	(Xie et al. 2012)
miRNA-143	↑	FNDC3B	Metastasis (+)	(Zhang et al. 2009)
miRNA-29a	↑	PTEN	Migration (+)	(Kong et al. 2011)
miRNA-21	↑	PTEN, PDCD4	Proliferation (−), Apoptosis (+),	(Connolly et al. 2008; Qiu et al. 2013:4; Liu et al. 2010)
miRNA-18a	↑	ERα	Proliferation (+)	(Liu et al. 2009)
miRNA-17-92	↑	E2F1	Proliferation (−), Anchorage-independent growth (−)	(Connolly et al. 2008; Aguda et al. 2008)

TABLE 8.3
Deregulated miRNAs Involved in HCV-Induced HCC

miRNAs	Expression	Target	References
miRNA-30d	↑	GNAI2	(Yao et al. 2010)
miRNA-17/92	↑	HSP-27	(Yang et al. 2010)
miRNA-21	↑	PTEN, SPRY2, PDCD4, RHOB, MASPIN	(Zhu et al. 2008; Meng et al. 2007; Sayed et al. 2008:2; Edwin et al. 2006; Asangani et al. 2008:4)
miRNA-221	↑	CDKN1B/p27, CDKN1C/ p57, BMF	(Fornari et al. 2008b:1; Gramantieri et al. 2009; Tannapfel et al. 2000)
miRNA-222	↑	PPP2R2A	(Wong et al. 2010)
Let-7g	↑	C-myc, p16INK4A, COL1A2	(Ji et al. 2010:1; Lan et al. 2011)
miRNA-101	↑	MCL-1, FOS	(Su et al. 2009)
miRNA-122	↓	ADAM17	(Tsai et al. 2009)
miRNA-139	↓	ROCK2	(Wong et al. 2011)
miRNA-29	↓	BCL-2, MCL-1	(Xiong et al. 2010)

HCV modulates several miRNAs to facilitate hepatocyte growth towards tumorigenesis by regulating various signaling pathways. The expression of miRNA-181c was found to be reduced in the HCV infected hepatocytes, which was responsible for promoting hepatocyte growth (Mukherjee et al. 2014). Expression of miRNA-155 level was increased in patients infected with HCV, which was responsible for promoting the cell proliferation and tumorigenesis via activation β-catenin, cyclin D1 and, c-myc (Yiliang Zhang et al. 2012). In addition, miRNA-141 has been reported to enhance viral replication and tumorigenesis in HCV-infected primary human hepatocytes (Banaudha et al. 2011).

8.5   miRNAs IN CELL-CYCLE REGULATION

Unrestricted cellular proliferation and prolonged survival play critical roles in the process of hepatocarcinogenesis. Deregulated miRNAs have been shown to alter the functionality of several key cell cycle regulators like cyclins, cyclin-dependent kinases (CDKs), CDK inhibitors and RB1 (Retino-blastoma). Some of the dysregulated miRNAs known to alter cell-cycle functionality and contribute toward HCC progression include miR-122, miR-26a, miR-221, miR-222, miR-195, miR-34a, miR-193b, miR-519d, miR-138, and miR-15a/16-1. Among these miRNAs, the expression of miRNA-221, -222, -519d and -15a/16-1 goes up while miRNAs -26a, -34a, -122, -138, and -195 are downregulated in HCC. MiR-519d directly targets PTEN, AKT3 CDKN1A/p21, and TIMP2 in HCC cells while miR-221 and miR-222 target CDKN1B/p27/Kip1 and CDKN1C/p57/Kip2, and thus, facilitate cell proliferation and invasion while modulating the signaling pathways leading towards apoptosis (Fornari et al. 2012; Gramantieri et al. 2009). miRNA-138 targets cyclin-D3, while miRNA-122 has been demonstrated to target cyclin-G1 in HCC, and thus, inhibit tumor growth (B et al. 2015; Gramantieri et al. 2007). miRNA-34a, miRNA216a, and miRNA-195 have been demonstrated to arrest cells at the G1 phase of cell cycle by directly targeting cyclin D1 (Kota et al. 2009; Xu et al. 2013; Chen et al. 2012). In addition to this, miRNA-34a targets CDK2 and CDK4, whereas miRNA-26a and -195 target CDK6 and E2F3.

8.5.1 REGULATION OF APOPTOSIS

Aberrantly expressed miRNAs are known to modulate the signaling pathways leading towards apoptosis in HCC. Bcl2, a member of the anti-apoptotic protein family, is targeted by miRNA-15a/16-1 as well as miRNA-224 (Zhang et al. 2013). Studies suggest that HBV facilitates upregulation of anti-apoptotic protein Bcl2 by sequestering miRNA-15A/16-1, which in turn promotes chronic HBV infection (Liu et al. 2013). miRNA-221 is downregulated in HCC and has been demonstrated to inhibit apoptosis by controlling its target gene Bmf, a member of pro-apoptotic protein family (Gramantieri et al. 2009). The miRNA-106b-25 cluster has been demonstrated to target pro-apoptotic protein Bim, and its expression is inversely correlated with Bim gene expression in HCC (Li et al. 2009). Anti-apoptotic proteins Bcl2, Bcl-w and Mcl-1 are targeted by miRNA-122 and miRNA-29 (Gebeshuber et al. 2009).

8.5.2 EPITHELIAL–MESENCHYMAL TRANSITION (EMT)

The miRNA-200 family which includes miRNA-200a, miRNA-200b, miRNA-200c as well as miRNA-141, miRNA-429, along with miRNA-205, has been demonstrated to play a key role in modulation of the EMT pathway, metastasis and invasion by targeting transcriptional repressors ZEB1 and ZEB2 (Bracken et al. 2008). E-box and Z-box elements present at the promoter region of the pri-miRNA-200 family have conserved sequences for ZEB1 binding (Mizuguchi et al. 2014). ZEB1 binding at the E-box and Z-box element of pri-miRNA-200 results in reduced miRNA-200 transcription. ZEB1 mediated transcriptional suppression of miRNA-200 leads to an increased level of ZEB1 through less binding to its 3' UTR (Mizuguchi et al. 2014). miRNA-101 expression has been inversely correlated with the epithelial-mesenchymal transition phenomenon in HCC cells. ZEB1 3'UTR is a direct target for miRNA-101 which in turn suppresses the TGF-β mediated EMT transition in HCC (Zhao et al. 2015). Downregulation of miRNA-124 and miRNA-139 leads to increased metastatic and invasive potential of HCC cells. Studies demonstrated that miRNA-124 and miRNA-139 targeted ROCK2 and inhibited EMT and prognosis in HCC (Zheng et al. 2012; Wong et al. 2011).

8.5.3 ANGIOGENESIS

Angiogenesis occupies the key position in the process of progression to HCC and various miRNAs, including miRNA-195, miRNA-26a, miRNA-221/222, miRNA-29b, have been demonstrated to regulate the various aspects of angiogenesis. Downregulation of miRNA-195 leads to increased levels of vascular endothelial growth factor (VEGF) in tumor cells and promotes angiogenesis in endothelial cells by activating VEGFR-2 signaling (Wang et al. 2013). miRNA-26a mediated inhibition of PIK3C2α, and its downstream signaling pathway, as well as hepatocyte growth factor receptor (cMet), causes decreased production of VEGF-A in HCC cells which in turn impairs the VEGFR-2 signaling pathway in endothelial cells and suppresses angiogenesis (Yang et al. 2014; Chai et al. 2013). miRNA-29b mediated suppression of MMP-2 expression in HCC cells leads to impaired VEGF receptor-2 signaling in endothelial cells, and thus, exerts its anti-angiogenic function (Fang et al. 2011). Activation of the hepatoma derived growth factor paracrine pathway in HCC was shown to be associated with downregulation of miRNA-214, and thus, promoted tumor angiogenesis (Shih et al. 2012). Loss of miRNA-122 expression in HCC patients is directly correlated with the presence of metastasis. miRNA-122 mediated inhibition of tumor necrosis factor-α converting enzyme (TACE) suppresses angiogenesis and metastasis (Tsai et al. 2009). Modulation of downstream targets such as cyclin G1, p27Kip1, p57Kip2, and c-kit by miRNA-221 and miRNA-222 has been demonstrated to impact the angiogenic properties as well as proliferation and migration of endothelial cells (Poliseno et al. 2006). The MiR-15a-16-1 cluster targets Bcl2, cyclin D1, AKT serine/threonine protein kinase (AKT3), mitogen activated protein kinase, and NF-kappaB activator MAP3-KIP3, and thus, inhibits VEGF expression, cell proliferation and promotes apoptosis (Bonci et al. 2008).

8.5.4 INVASION AND METASTASIS

Deregulated miRNAs have a significant impact on advanced tumor features, including invasion and metastasis. Many signaling pathways involving the HGF/MET axis have been demonstrated to confer invasive potential to HCC cells. The transmembrane tyrosine kinase receptor (MET) binds to its ligand, hepatocyte growth factor (HGF) and facilitates hepatocyte survival, angiogenesis, proliferation and migration (Birchmeier et al. 2003; Benvenuti and Comoglio 2007). miRNA-1, miRNA-23b, miRNA-34a and miRNA-199-3phave been shown to regulate the expression of MET (Fornari et al. 2010; Datta et al. 2008). In 40–70% of HCC cases, MET expression has been shown to be upregulated via downregulation of miRNAs-1, -23b, -34a, and -199-3p (Boix et al. 1994; Suzuki et al. 1994).

8.6 CLINICAL IMPLICATIONS OF THE miRNAs

8.6.1 miRNAs IN DIAGNOSTICS

miRNA expression profiling could be a prognostic and diagnostic tool for classification purposes in HCC. Many previous studies and reports have summarized the potential applications of miRNAs as diagnostic and prognostic markers in many cancers, including in liver cancer (Pant and Venugopal 2017; Callegari et al. 2013; Aravalli 2013). Tables 8.1 through 8.3 summarize the currently available information about several miRNAs and their potential prognostic significance in HCC. It is shown that miRNA-200c, miRNA-141 and miRNA-126, alone or in combination, could be used with high accuracy to distinguish primary HCC for classification purposes compared to the other tumor metastases in the liver. Moreover, the ratio between miRNA-205 and miRNA-194 expression has also been shown to be useful in differentiating gastrointestinal tumors (Callegari et al. 2013), in which the liver is the major metastatic site for these tumors.

This recent area of research also focuses on the potential of miRNAs and their applications in circulating biomarkers for HCC. This is because researchers find different levels of miRNAs in the serum, plasma and tissues of patients affected by a range of liver diseases in comparison with healthy subjects (Pant and Venugopal 2017; Li et al. 2014). The levels of miR-122 and miR-21 have been reported by more than one study to be significantly higher in patients with HCC. And, although this needs further verification through additional studies, their increased levels were also detected in chronic hepatitis patients, indicating their efficacy as clinical HCC markers (Gramantieri et al. 2008). These properties of miRNAs as potential biomarkers could improve our capability to develop non-invasive diagnostic biomarkers for HCC. Nevertheless, at present, further research is still necessary prior to utilizing miRNAs in the clinic as biomarkers for HCC.

8.6.2 MIRNAS IN HCC THERAPEUTICS

Apart from being used as diagnostic biomarkers for HCC, miRNAs have also been used as potential therapeutic molecules for the treatment of many cancers, including HCC. In recent years, several studies have demonstrated that strategies based on miRNA activity modulation could be used as an approach to treating liver cancer (Wang and Chen 2014). There are mainly two types of strategies to treat cancer by using miRNAs which are described below.

8.6.2.1 Inhibition of miRNAs

Inhibition of miRNA-221 by intratumoral injections of anti-miR-221 was shown to have antitumor activity in prostate carcinoma (Mercatelli et al. 2008). Application of anti-miR-221 reduced proliferation of tumor cells and tumor size and promoted survival rates in HCC mouse models as compared to untreated animals (Park et al. 2011). By using anti-miR-21 molecules, it led to the complete inhibition of lymphoid-like malignancies in mice (Medina et al. 2010). In addition, Anti-miR-21 was also reported to have significant antitumor activity in multiple myeloma xenografts mice models (Leone et al. 2013). The role of miR-122 in HCV-RNA accumulation is well studied. The treatment of anti-miR-122 oligonucleotide demonstrated the long-lasting suppression of virus accumulation in chronically HCV-infected nonhuman primates (Hildebrandt-Eriksen et al. 2012). Miravirsen SPC3649 is the first drug targeting miRNAs that has been used in the treatment of HCV infection; it is currently under Phase II of a clinical trial (Lindow and Kauppinen 2012).

8.6.2.2 Restoration of miRNAs

Another approach to treat cancer is based on substitution of miRNAs which can suppress the tumor. The over expression of miRNA-26a and miRNA-122 using adenovirus associated vector system inhibited tumorigenicity in HCC mouse models (Kota et al. 2009; Hsu et al. 2012). Delivery of miRNA-199 using an adenovirus associated vector system effectively reduced the tumor size in HCC xenografts mice. Delivery of miR-29bin nude mouse xenograft models successfully sensitized HCC cell progression and size via inducing various apoptotic signals in tumor cells (Xiong et al. 2010). One study showed that treatment to miR-34a with the cytokine interleukin-24 manifested synergistic antitumor effects in a xenograft mice model of HCC (Lou et al. 2013), indicating the possible role of miRNAs in HCC therapy.

8.7 CONCLUSION

The invention of abnormally expressed miRNAs in HCC has revealed some novel mechanisms involved in liver tumorigenesis. Understanding the molecular mechanism of tumorigenesis has opened up areas of progress for more rational classification and therapeutic approaches. Abnormally expressed miRNAs can be associated with pathological and clinical features of HCC. Understanding this can make miRNA expression a potentially useful tool for HCC classification and prognostic stratification in early HCC, where the availability of potentially curative approaches requires a more sophisticated diagnostic approach. Finally, by revealing which miRNAs are up- or downregulated in liver tumorigenesis, new potential therapeutic targets have been revealed. While miRNA-based approaches are not presently used in the clinic, their potential applications are expanding in various areas of interest. Prognostic stratification, follow-up monitoring, and innovative therapeutic approaches are areas that might benefit from the use of miRNAs. It should be noted that it has only been during the last 10 years that investigation of miRNAs in cancer has begun, and many more studies are needed to move this field forward into the clinical setting. Validation based on prospective studies of the use of miRNAs as cancer biomarkers is needed. The application of miRNA replacement or inhibition approaches needs larger preclinical studies to assess the potential efficacy in specific contexts.

REFERENCES

Acharya, Subrat K. 2014. "Epidemiology of Hepatocellular Carcinoma in India." *Journal of Clinical and Experimental Hepatology* 4 (Suppl 3): S27–33.

Aguda, Baltazar D., Yangjin Kim, Melissa G. et al. 2008. "MicroRNA Regulation of a Cancer Network: Consequences of the Feedback Loops Involving miR-17-92, E2F, and Myc." *Proceedings of the National Academy of Sciences of the United States of America* 105 (50): 19678–83.

Ambros, Victor. 2004. "The Functions of Animal microRNAs." *Nature* 431 (7006): 350–55.

Aravalli, Rajagopal N. 2013. "Development of MicroRNA Therapeutics for Hepatocellular Carcinoma." *Diagnostics* 3 (1): 170–91.

Asangani, Irfan A., Rasheed, Sheikh AK., Nikolova K. Dimitringa et al. 2008. "MicroRNA-21 (miR-21) "Post-Transcriptionally Downregulates Tumor Suppressor Pdcd4 and Stimulates Invasion, Intravasation and Metastasis in Colorectal Cancer." *Oncogene* 27 (15): 2128–36.

Au, Sandy Leung-Kuen, Carmen Chak-Lui Wong, Joyce Man-Fong Lee et al. 2012. "Enhancer of Zeste Homolog 2 Epigenetically Silences Multiple Tumor Suppressor microRNAs to Promote Liver Cancer Metastasis." *Hepatology (Baltimore, Md.)* 56 (2): 622–31.

Baffy, György, Elizabeth M. Brunt, and Stephen H. Caldwell. 2012. "Hepatocellular Carcinoma in Non-Alcoholic Fatty Liver Disease: An Emerging Menace." *Journal of Hepatology* **56** (6): 1384–91.

Banaudha, Krishna, Michael Kaliszewski, Tamara Korolnek et al. 2011. "MicroRNA Silencing of Tumor Suppressor DLC-1 Promotes Efficient Hepatitis C Virus Replication in Primary Human Hepatocytes." *Hepatology (Baltimore, Md.)* **53** (1): 53–61.

Bandyopadhyay, Sarmistha, Robin C. Friedman, Rebecca T. Marquez et al. 2011. "Hepatitis C Virus Infection and Hepatic Stellate Cell Activation Downregulate miR-29: miR-29 Overexpression Reduces Hepatitis C Viral Abundance in Culture." *The Journal of Infectious Diseases* **203** (12): 1753–62.

Bao, Longlong, Yan Yan, Can Xu et al. 2013. "MicroRNA-21 Suppresses PTEN and hSulf-1 Expression and Promotes Hepatocellular Carcinoma Progression through AKT/ERK Pathways." *Cancer Letters* **337** (2): 226–36.

Benvenuti, Silvia, and Paolo M. Comoglio. 2007. "The MET Receptor Tyrosine Kinase in Invasion and Metastasis." *Journal of Cellular Physiology* **213** (2): 316–25.

Birchmeier, Carmen, Walter Birchmeier, Ermanno Gherardi, 2003. "Met, Metastasis, Motility and More." *Nature Reviews. Molecular Cell Biology* **4** (12): 915–25.

Boix, L., J. L. Rosa, F. Ventura et al. 1994. "C-Met mRNA Overexpression in Human Hepatocellular Carcinoma." *Hepatology (Baltimore, Md.)* **19** (1): 88–91.

Bonci, Désirée, Valeria Coppola, Maria Musumeci et al. 2008. "The miR-15a–miR-16-1 Cluster Controls Prostate Cancer by Targeting Multiple Oncogenic Activities." *Nature Medicine* **14** (11): 1271–77.

Bracken, Cameron P., Philip A. Gregory, Natasha Kolesnikoff et al. 2008. "A Double-Negative Feedback Loop between ZEB1-SIP1 and the microRNA-200 Family Regulates Epithelial-Mesenchymal Transition." *Cancer Research* **68** (19): 7846–54.

Buurman, Reena, Engin Gürlevik, Vera Schäffer et al. 2012. "Histone Deacetylases Activate Hepatocyte Growth Factor Signaling by Repressing microRNA-449 in Hepatocellular Carcinoma Cells." *Gastroenterology* **143** (3): 811-820.e1-15.

Callegari, Elisa, Bahaeldin K. Elamin, Silvia Sabbioni et al. 2013. "Role of microRNAs in Hepatocellular Carcinoma: A Clinical Perspective." *OncoTargets and Therapy* **6** (September): 1167–78.

Chai, Zong-Tao, Jian Kong, Xiao-Dong Zhu et al. 2013. "MicroRNA-26a Inhibits Angiogenesis by Down-Regulating VEGFA through the PIK3C2α/Akt/HIF-1α Pathway in Hepatocellular Carcinoma." *PLOS ONE* **8** (10): e77957.

Chen, Lizao, Jianming Zheng, Yan Zhang et al. 2011. "Tumor-Specific Expression of microRNA-26a Suppresses Human Hepatocellular Carcinoma Growth via Cyclin-Dependent and -Independent Pathways." *Molecular Therapy: The Journal of the American Society of Gene Therapy* **19** (8): 1521–28.

Chen, Po-Jen, Shiou-Hwei Yeh, Wan-Hsin Liu et al. 2012. "Androgen Pathway Stimulates MicroRNA-216a Transcription to Suppress the Tumor Suppressor in Lung Cancer-1 Gene in Early Hepatocarcinogenesis." *Hepatology* **56** (2): 632–43.

Chen, Yanni, Ao Shen, Paul J. Rider et al. 2011. "A Liver-Specific microRNA Binds to a Highly Conserved RNA Sequence of Hepatitis B Virus and Negatively Regulates Viral Gene Expression and Replication." *FASEB Journal: Official Publication of the Federation of American Societies for Experimental Biology* **25** (12): 4511–21.

Cheng, Ju-Chien, Yung-Ju Yeh, Ching-Ping Tseng et al. 2012. "Let-7b Is a Novel Regulator of Hepatitis C Virus Replication." *Cellular and Molecular Life Sciences: CMLS* **69** (15): 2621–33.

Cheng, Jun, Lin Zhou, Qin-Fen Xie et al. 2010. "The Impact of miR-34a on Protein Output in Hepatocellular Carcinoma HepG2 Cells." *Proteomics* **10** (8): 1557–72.

Connolly, Erin C., Koenraad Van Doorslaer, Leslie E. Rogler et al. 2010. "Overexpression of miR-21 Promotes an in Vitro Metastatic Phenotype by Targeting the Tumor Suppressor RHOB." *Molecular Cancer Research* **8** (5): 691–700.

Connolly, Erin, Margherita Melegari, Pablo Landgraf et al. 2008. "Elevated Expression of the miR-17-92 Polycistron and miR-21 in Hepadnavirus-Associated Hepatocellular Carcinoma Contributes to the Malignant Phenotype." *The American Journal of Pathology* **173** (3): 856–64.

Conrad, K. Dominik, Florian Giering, Corinna Erfurth et al. 2013. "MicroRNA-122 Dependent Binding of Ago2 Protein to Hepatitis C Virus RNA Is Associated with Enhanced RNA Stability and Translation Stimulation." *PloS One* **8** (2): e56272.

Cougot, Delphine, Christine Neuveut, and Marie Annick Buendia. 2005. "HBV Induced Carcinogenesis." *Journal of Clinical Virology* **34** Suppl 1: S75–78.

Coulouarn, C., V. M. Factor, J. B. Andersen et al. 2009. "Loss of miR-122 Expression in Liver Cancer Correlates with Suppression of the Hepatic Phenotype and Gain of Metastatic Properties." *Oncogene* **28** (40): 3526–36.

Datta, Jharna, Huban Kutay, Mohd W. Nasser et al. 2008. "Methylation Mediated Silencing of MicroRNA-1 Gene and Its Role in Hepatocellular Carcinogenesis." *Cancer Research* **68** (13): 5049–58.

Dhayat, Sameer A., Wolf A. Mardin, Gabriele Köhler et al. 2014. "The microRNA-200 Family—A Potential Diagnostic Marker in Hepatocellular Carcinoma?" *Journal of Surgical Oncology* **110** (4): 430–38.

Di Bisceglie, Adrian M. 2009. "Hepatitis B And Hepatocellular Carcinoma." *Hepatology (Baltimore, Md.)* **49** (5 Suppl): S56–60.

Di Fazio, Pietro, Roberta Montalbano, Daniel Neureiter et al. 2012. "Downregulation of HMGA2 by the Pan-Deacetylase Inhibitor Panobinostat Is Dependent on Hsa-Let-7b Expression in Liver Cancer Cell Lines." *Experimental Cell Research* **318** (15): 1832–43.

Edwin, Francis, Rakesh Singh, Raelene Endersby et al. 2006. "The Tumor Suppressor PTEN Is Necessary for Human Sprouty 2-Mediated Inhibition of Cell Proliferation." *The Journal of Biological Chemistry* **281** (8): 4816–22.

El-Serag, Hashem B. 2002. "Hepatocellular Carcinoma: An Epidemiologic View." *Journal of Clinical Gastroenterology* **35** (5 Suppl 2): S72–78.

El-Serag, Hashem B. 2004. "Hepatocellular Carcinoma: Recent Trends in the United States." *Gastroenterology* **127** (5 Suppl 1): S27–34.

Fan, Qin, Minyi He, Xinjun Deng et al. 2013. "Derepression of c-Fos Caused by microRNA-139 down-Regulation Contributes to the Metastasis of Human Hepatocellular Carcinoma." *Cell Biochemistry and Function* **31** (4): 319–24.

Fang, Jian-Hong, Hui-Chao Zhou, Chunxian Zeng et al. 2011. "MicroRNA-29b Suppresses Tumor Angiogenesis, Invasion, and Metastasis by Regulating Matrix Metalloproteinase 2 Expression." *Hepatology (Baltimore, Md.)* **54** (5): 1729–40.

Fang, YuXiang, Jing-Lun Xue, Qi Shen et al. 2012. "MicroRNA-7 Inhibits Tumor Growth and Metastasis by Targeting the Phosphoinositide 3-Kinase/Akt Pathway in Hepatocellular Carcinoma." *Hepatology (Baltimore, Md.)* **55** (6): 1852–62.

Fattovich, Giovanna, Tommaso Stroffolini, Irene Zagni, and Francesco Donato. 2004. "Hepatocellular Carcinoma in Cirrhosis: Incidence and Risk Factors." *Gastroenterology* **127** (5 Suppl 1): S35-50.

Fornari, Francesca, Laura Gramantieri, Manuela Ferracin et al. 2008. "MiR-221 Controls CDKN1C/p57 and CDKN1B/p27 Expression in Human Hepatocellular Carcinoma." *Oncogene* **27** (43): 5651–61. doi:10.1038/onc.2008.178.

Fornari, Francesca, Maddalena Milazzo, Pasquale Chieco et al. 2010. "MiR-199a-3p Regulates mTOR and c-Met to Influence the Doxorubicin Sensitivity of Human Hepatocarcinoma Cells." *Cancer Research* **70** (12): 5184–93.

Fornari, Francesca, Maddalena Milazzo, Pasquale Chieco et al. 2012. "In Hepatocellular Carcinoma miR-519d Is up-Regulated by p53 and DNA Hypomethylation and Targets CDKN1A/p21, PTEN, AKT3 and TIMP2." *The Journal of Pathology* **227** (3): 275–85.

Fujita, Shuji, Taiji Ito, Taketoshi Mizutani et al. 2008. "miR-21 Gene Expression Triggered by AP-1 Is Sustained through a Double-Negative Feedback Mechanism." *Journal of Molecular Biology* **378** (3): 492–504.

Garofalo, Michela, Gianpiero Di Leva, Giulia Romano et al. 2009. "miR-221&222 Regulate TRAIL Resistance and Enhance Tumorigenicity through PTEN and TIMP3 Downregulation." *Cancer Cell* **16** (6): 498–509.

Gastaldi, Giacomo, Nicolas Goossens, Sophie Clément et al. 2017. "Current Level of Evidence on Causal Association between Hepatitis C Virus and Type 2 Diabetes: A Review." *Journal of Advanced Research* **8** (2): 149–59.

Gebeshuber, Christoph A., Kurt Zatloukal, and Javier Martinez. 2009. "miR-29a Suppresses Tristetraprolin, Which Is a Regulator of Epithelial Polarity and Metastasis." *EMBO Reports* **10** (4): 400–5.

Goh, George Boon-Bee, Pik-Eu Chang, and Chee-Kiat Tan. 2015. "Changing Epidemiology of Hepatocellular Carcinoma in Asia." *Best Practice & Research. Clinical Gastroenterology* **29** (6): 919–28.

Gomaa, Asmaa Ibrahim, Shahid A Khan, Mireille B Toledano et al. 2008. "Hepatocellular Carcinoma: Epidemiology, Risk Factors and Pathogenesis." *World Journal of Gastroenterology: WJG* **14** (27): 4300–08.

Gramantieri, Laura, Manuela Ferracin, Francesca Fornari et al. 2007. "Cyclin G1 Is a Target of miR-122a, a microRNA Frequently down-Regulated in Human Hepatocellular Carcinoma." *Cancer Research* **67** (13): 6092–99.

Gramantieri, Laura, Francesca Fornari, Elisa Callegari et al. 2008. "MicroRNA Involvement in Hepatocellular Carcinoma." *Journal of Cellular and Molecular Medicine* **12** (6A): 2189–204.

Gramantieri, Laura, Francesca Fornari, Manuela Ferracin et al. 2009. "MicroRNA-221 Targets Bmf in Hepatocellular Carcinoma and Correlates with Tumor Multifocality." *Clinical Cancer Research: An Official Journal of the American Association for Cancer Research* **15** (16): 5073–81.

Guo, Hongyan, Haiying Liu, Keith Mitchelson et al. 2011. "MicroRNAs-372/373 Promote the Expression of Hepatitis B Virus through the Targeting of Nuclear Factor I/B." *Hepatology (Baltimore, Md.)* **54** (3): 808–19.

Han, Han, Dan Sun, Wenjuan Li, Hongxing Shen et al. 2013. "A c-Myc-MicroRNA Functional Feedback Loop Affects Hepatocarcinogenesis." *Hepatology (Baltimore, Md.)* **57** (6): 2378–89.

Hayes, C. Nelson, Sakura Akamatsu, Masataka Tsuge et al. 2012. "Hepatitis B Virus-Specific miRNAs and Argonaute2 Play a Role in the Viral Life Cycle." *PLoS One* **7** (10): e47490.

Hildebrandt-Eriksen, Elisabeth S., Vibeke Aarup et al. 2012. "A Locked Nucleic Acid Oligonucleotide Targeting microRNA 122 Is Well-Tolerated in Cynomolgus Monkeys." *Nucleic Acid Therapeutics* **22** (3): 152–61.

Hoshida, Yujin, Bryan C. Fuchs, Nabeel Bardeesy et al. 2014. "Pathogenesis and Prevention of Hepatitis C Virus-Induced Hepatocellular Carcinoma." *Journal of Hepatology* **61** (1 Suppl): S79–90.

Hou, Jin, Li Lin, Weiping Zhou et al. 2011. "Identification of miRNomes in Human Liver and Hepatocellular Carcinoma Reveals miR-199a/B-3p as Therapeutic Target for Hepatocellular Carcinoma." *Cancer Cell* **19** (2): 232–43.

Huang Bo, Huiwen Li, Liyu Huang et al. 2015. "Clinical Significance of microRNA 138 and Cyclin D3 in Hepatocellular Carcinoma." *The Journal of Surgical Research* **193** (2): 718–23.

Hsu, Shu-Hao, Bo Wang, Janaiah Kota et al. "Essential Metabolic, Anti-Inflammatory, and Anti-Tumorigenic Functions of miR-122 in Liver." *The Journal of Clinical Investigation* **122** (8): 2871–83.

Hu, Jun, Yaxing Xu, Junli Hao et al. 2012. "MiR-122 in Hepatic Function and Liver Diseases." *Protein & Cell* **3** (5): 364–71.

Hu, Wei, Xuejun Wang, Xiaoran Ding et al. 2012. "MicroRNA-141 Represses HBV Replication by Targeting PPARA." *PLoS One* **7** (3): e34165.

Huang, Jinfeng, Yue Wang, Yingjun Guo, and Shuhan Sun. 2010. "Down-Regulated microRNA-152 Induces Aberrant DNA Methylation in Hepatitis B Virus-Related Hepatocellular Carcinoma by Targeting DNA Methyltransferase 1." *Hepatology (Baltimore, Md.)* **52** (1): 60–70.

Jangra, Rohit K., Minkyung Yi, and Stanley M. Lemon. 2010. "Regulation of Hepatitis C Virus Translation and Infectious Virus Production by the microRNA miR-122." *Journal of Virology* **84** (13): 6615–25.

Ji, Junfang, Lei Zhao, Anuradha Budhu et al. 2010. "Let-7g Targets Collagen Type I alpha2 and Inhibits Cell Migration in Hepatocellular Carcinoma." *Journal of Hepatology* **52** (5): 690–97.

Jiang, Runqiu, Lei Deng, Liang Zhao et al. 2011. "miR-22 Promotes HBV-Related Hepatocellular Carcinoma Development in Males." *Clinical Cancer Research: An Official Journal of the American Association for Cancer Research* **17** (17): 5593–5603.

Jin, Jiang, Shanhong Tang, Lin Xia et al. 2013. "MicroRNA-501 Promotes HBV Replication by Targeting HBXIP." *Biochemical and Biophysical Research Communications* **430** (4): 1228–33.

Johnson, Steven M., Helge Grosshans, Jaclyn Shingara et al. 2005. "RAS Is Regulated by the Let-7 microRNA Family." *Cell* **120** (5): 635–47.

Jopling, Catherine. 2012. "Liver-Specific microRNA-122: Biogenesis and Function." *RNA Biology* **9** (2): 137–42.

Jung, Yong Jin, Jin-Wook Kim, Soo Jin Park et al. 2013. "C-Myc-Mediated Overexpression of miR-17-92 Suppresses Replication of Hepatitis B Virus in Human Hepatoma Cells." *Journal of Medical Virology* **85** (6): 969–78.

Kar, Premashis. 2014. "Risk Factors for Hepatocellular Carcinoma in India." *Journal of Clinical and Experimental Hepatology* **4** (Suppl 3): S34–42.

Kasahara, Kotaro, Takahiro Taguchi, Ichiro Yamasaki et al. 2002. "Detection of Genetic Alterations in Advanced Prostate Cancer by Comparative Genomic Hybridization." *Cancer Genetics and Cytogenetics* **137** (1): 59–63.

Kew, Michael C. 2013. "Aflatoxins as a Cause of Hepatocellular Carcinoma." *Journal of Gastrointestinal and Liver Diseases: JGLD* **22** (3): 305–10.

Kong, Guangyao, Junping Zhang, Shuai Zhang et al. 2011. "Upregulated MicroRNA-29a by Hepatitis B Virus X Protein Enhances Hepatoma Cell Migration by Targeting PTEN in Cell Culture Model." *PLoS One* **6** (5): e19518.

Kong, William, Hua Yang, Lili He et al. 2008. "MicroRNA-155 Is Regulated by the Transforming Growth Factor Beta/Smad Pathway and Contributes to Epithelial Cell Plasticity by Targeting RhoA." *Molecular and Cellular Biology* **28** (22): 6773–84.

Kota, Janaiah, Raghu R. Chivukula, Kathryn A. O'Donnell et al. 2009. "Therapeutic microRNA Delivery Suppresses Tumorigenesis in a Murine Liver Cancer Model." *Cell* **137** (6): 1005–17.

Lan, Fei-Fei, Hua Wang, Yang-Chao Chen et al. 2011. "Hsa-Let-7g Inhibits Proliferation of Hepatocellular Carcinoma Cells by Downregulation of c-Myc and Upregulation of p16(INK4A)." *International Journal of Cancer* **128** (2): 319–31.

Lan, Sheng-Hui, Shan-Ying Wu, Roberto Zuchini et al. 2014. "Autophagy Suppresses Tumorigenesis of Hepatitis B Virus-Associated Hepatocellular Carcinoma through Degradation of microRNA-224." *Hepatology (Baltimore, Md.)* **59** (2): 505–17.

Leone, Emanuela, Eugenio Morelli, Maria T. Di Martino et al. 2013. "Targeting miR-21 Inhibits in Vitro and in Vivo Multiple Myeloma Cell Growth." *Clinical Cancer Research: An Official Journal of the American Association for Cancer Research* **19** (8): 2096–106.

Li, Changfei, Yanzhong Wang, Saifeng Wang et al. 2013. "Hepatitis B Virus mRNA-Mediated miR-122 Inhibition Upregulates PTTG1-Binding Protein, Which Promotes Hepatocellular Carcinoma Tumor Growth and Cell Invasion." *Journal of Virology* **87** (4): 2193–205.

Li, Dong, Xingguang Liu, Li Lin et al. 2011. "MicroRNA-99a Inhibits Hepatocellular Carcinoma Growth and Correlates with Prognosis of Patients with Hepatocellular Carcinoma." *The Journal of Biological Chemistry* **286** (42): 36677–85.

Li, Dong, Pengyuan Yang, Hua Li et al. 2012. "MicroRNA-1 Inhibits Proliferation of Hepatocarcinoma Cells by Targeting Endothelin-1." *Life Sciences* **91** (11–12): 440–7.

Li, Gang, Guohong Cai, Demin Li et al. 2014. "MicroRNAs and Liver Disease: Viral Hepatitis, Liver Fibrosis and Hepatocellular Carcinoma." *Postgraduate Medical Journal* **90** (1060): 106–12.

Li, Na, Hanjiang Fu, Yi Tie et al. 2009. "miR-34a Inhibits Migration and Invasion by Down-regulation of c-Met Expression in Human Hepatocellular Carcinoma Cells." *Cancer Letters* **275** (1): 44–53.

Li, Shuai, Hanjiang Fu, Yulan Wang et al. 2009. "MicroRNA-101 Regulates Expression of the v-Fos FBJ Murine Osteosarcoma Viral Oncogene Homolog (FOS) Oncogene in Human Hepatocellular Carcinoma." *Hepatology (Baltimore, Md.)* **49** (4): 1194–202.

Li, Xiaofei, Wenjun Yang, Weiwei Ye et al. 2014. "microRNAs: Novel Players in Hepatitis C Virus Infection." *Clinics and Research in Hepatology and Gastroenterology* **38** (6): 664–75.

Li, Yang, Weiqi Tan, Thomas W. L. Neo et al. 2009. "Role of the miR-106b-25 microRNA Cluster in Hepatocellular Carcinoma." *Cancer Science* **100** (7): 1234–42.

Lin, Cliff Ji-Fan, Hong-Yi Gong, Hung-Chia Tseng et al. 2008. "miR-122 Targets an Anti-Apoptotic Gene, Bcl-W, in Human Hepatocellular Carcinoma Cell Lines." *Biochemical and Biophysical Research Communications* **375** (3): 315–20.

Lindow, Morten, and Sakari Kauppinen. 2012. "Discovering the First microRNA-Targeted Drug." *The Journal of Cell Biology* **199** (3): 407–12.

Liu, Changzheng, Jia Yu, Shuangni Yu et al. 2010. "MicroRNA-21 Acts as an Oncomir through Multiple Targets in Human Hepatocellular Carcinoma." *Journal of Hepatology* **53** (1): 98–107.

Liu, Ningning, Jinfang Zhang, Tong Jiao et al. 2013. "Hepatitis B Virus Inhibits Apoptosis of Hepatoma Cells by Sponging the MicroRNA 15a/16 Cluster." *Journal of Virology* **87** (24): 13370–8.

Liu, Shupeng, Weixing Guo, Jie Shi et al. 2012. "MicroRNA-135a Contributes to the Development of Portal Vein Tumor Thrombus by Promoting Metastasis in Hepatocellular Carcinoma." *Journal of Hepatology* **56** (2): 389–96.

Liu, Wan-Hsin, Shiou-Hwei Yeh, Cho-Chun Lu et al. 2009. "MicroRNA-18a Prevents Estrogen Receptor-α Expression, Promoting Proliferation of Hepatocellular Carcinoma Cells." *Gastroenterology* **136** (2): 683–93.

Lou, Wenjia, Qing Chen, Leina Ma et al. 2013. "Oncolytic Adenovirus Co-Expressing miRNA-34a and IL-24 Induces Superior Antitumor Activity in Experimental Tumor Model." *Journal of Molecular Medicine (Berlin, Germany)* **91** (6): 715–25.

Margini, Cristina, and Jean F. Dufour. 2016. "The Story of HCC in NAFLD: From Epidemiology, across Pathogenesis, to Prevention and Treatment." *Liver International: Official Journal of the International Association for the Study of the Liver* **36** (3): 317–24.

Medina, Pedro P., Mona Nolde, and Frank J. Slack. 2010. "OncomiR Addiction in an in Vivo Model of microRNA-21-Induced Pre-B-Cell Lymphoma." *Nature* **467** (7311): 86–90.

Meng, Fanyin, Roger Henson, Hania Wehbe-Janek et al. 2007. "MicroRNA-21 Regulates Expression of the PTEN Tumor Suppressor Gene in Human Hepatocellular Cancer." *Gastroenterology* **133** (2): 647–58.

Mercatelli, Neri, Valeria Coppola, Desirée Bonci et al. 2008. "The Inhibition of the Highly Expressed miR-221 and miR-222 Impairs the Growth of Prostate Carcinoma Xenografts in Mice." *PloS One* **3** (12): e4029.

Mercer, K. E., L. Hennings, and M. J. J. Ronis. 2015. "Alcohol Consumption, Wnt/β-Catenin Signaling, and Hepatocarcinogenesis." *Advances in Experimental Medicine and Biology* **815**: 185–95.

Mizuguchi, Yoshiaki, Kumiko Isse, Susan Specht et al. 2014. "Small Proline Rich Protein 2a in Benign and Malignant Liver Disease." *Hepatology (Baltimore, Md.)* **59** (3): 1130–43.

Mukherjee, Anupam, Shubham Shrivastava, Joydip Bhanja Chowdhury et al. 2014. "Transcriptional Suppression of miR-181c by Hepatitis C Virus Enhances Homeobox A1 Expression." *Journal of Virology* **88** (14): 7929–40.

Murakami, Yoshiki, Taeko Yasuda, Kenichi Saigo et al. 2006. "Comprehensive Analysis of microRNA Expression Patterns in Hepatocellular Carcinoma and Non-Tumorous Tissues." *Oncogene* **25** (17): 2537–45.

Noh, Ji Heon, Young Gyoon Chang, Min Gyu Kim et al. 2013. "MiR-145 Functions as a Tumor Suppressor by Directly Targeting Histone Deacetylase 2 in Liver Cancer." *Cancer Letters* **335** (2): 455–62.

Novellino, Luisa, Riccardo L. Rossi, Ferruccio Bonino et al. 2012. "Circulating Hepatitis B Surface Antigen Particles Carry Hepatocellular microRNAs." *PloS One* **7** (3): e31952.

Oliveria Andrade, Luis Jesuino de, Argemiro D'Oliveira et al. 2009. "Association between Hepatitis C and Hepatocellular Carcinoma." *Journal of Global Infectious Diseases* **1** (1): 33–7.

Pan, Xuan, Zhao-Xia Wang, and Rui Wang. 2010. "MicroRNA-21: A Novel Therapeutic Target in Human Cancer." *Cancer Biology & Therapy* **10** (12): 1224–32.

Pant, Kishor, and Senthil K. Venugopal. 2017. "Circulating microRNAs: Possible Role as Non-Invasive Diagnostic Biomarkers in Liver Disease." *Clinics and Research in Hepatology and Gastroenterology*. **41** (4): 370–7.

Park, Jong-Kook, Takayuki Kogure, Gerard J. Nuovo et al. 2011. "miR-221 Silencing Blocks Hepatocellular Carcinoma and Promotes Survival." *Cancer Research* **71** (24): 7608–16.

Parkin, Donald Maxwell. 2006. "The Global Health Burden of Infection-Associated Cancers in the Year 2002." *International Journal of Cancer* **118** (12): 3030–44.

Pedersen, Irene M., Guofeng Cheng, Stefan Wieland et al. 2007. "Interferon Modulation of Cellular microRNAs as an Antiviral Mechanism." *Nature* **449** (7164): 919–22.

Poliseno, Laura, Andrea Tuccoli, Laura Mariani et al. 2006. "MicroRNAs Modulate the Angiogenic Properties of HUVECs." *Blood* **108** (9): 3068–71.

Potenza, Nicoletta, Umberto Papa, Nicola Mosca et al. 2011. "Human microRNA Hsa-miR-125a-5p Interferes with Expression of Hepatitis B Virus Surface Antigen." *Nucleic Acids Research* **39** (12): 5157–63.

Qiu, Xi, Song Dong, Fengchang Qiao et al. 2013. "HBx-Mediated miR-21 Upregulation Represses Tumor-Suppressor Function of PDCD4 in Hepatocellular Carcinoma." *Oncogene* **32** (27): 3296–3305.

Reinheart, Brenda J., Frank J. Slack, Michael Basson et al. 2000. "The 21-Nucleotide Let-7 RNA Regulates Developmental Timing in Caenorhabditis Elegans." *Nature* **403** (6772): 901–6.

Ribas, Judit, and Shawn E. Lupold. 2010. "The Transcriptional Regulation of miR-21, Its Multiple Transcripts, and Their Implication in Prostate Cancer." *Cell Cycle (Georgetown, Tex.)* 9 (5): 923–9.

Salvi, Alessandro, Cristiano Sabelli, Silvia Moncini et al. 2009. "MicroRNA-23b Mediates Urokinase and c-Met Downmodulation and a Decreased Migration of Human Hepatocellular Carcinoma Cells." *The FEBS Journal* **276** (11): 2966–82.

Santhekadur, Prasanna Kumar, Swadesh K. Das, Rachel Gredler et al. 2012. "Multifunction Protein Staphylococcal Nuclease Domain Containing 1 (SND1) Promotes Tumor Angiogenesis in Human Hepatocellular Carcinoma through Novel Pathway That Involves Nuclear Factor κB and miR-221." *The Journal of Biological Chemistry* **287** (17): 13952–8.

Sarma, Nayan J., Venkataswarup Tiriveedhi, Vijay Subramanian et al. 2012. "Hepatitis C Virus Mediated Changes in miRNA-449a Modulates Inflammatory Biomarker YKL40 through Components of the NOTCH Signaling Pathway." *PLOS ONE* **7** (11): e50826.

Sayed, Danish, Shweta Rane, Jacqueline Lypowy et al. 2008. "MicroRNA-21 Targets Sprouty2 and Promotes Cellular Outgrowths." *Molecular Biology of the Cell* **19** (8): 3272–82.

Scisciani, Cecilia, Stefania Vossio, Francesca Guerrieri et al. 2012. "Transcriptional Regulation of miR-224 Upregulated in Human HCCs by NFκB Inflammatory Pathways." *Journal of Hepatology* **56** (4): 855–61.

Shah, Yatrik M., Keiichirou Morimura, Qian Yang et al. 2007. "Peroxisome Proliferator-Activated Receptor Alpha Regulates a microRNA-Mediated Signaling Cascade Responsible for Hepatocellular Proliferation." *Molecular and Cellular Biology* **27** (12): 4238–47.

Shen, Qingli, Vito R. Cicinnati, Xiaoyong Zhang et al. 2010. "Role of microRNA-199a-5p and Discoidin Domain Receptor 1 in Human Hepatocellular Carcinoma Invasion." *Molecular Cancer* **9**: 227.

Shen, Qingyu, Hyun Jin Bae, Jung Woo Eun et al. 2014. "MiR-101 Functions as a Tumor Suppressor by Directly Targeting Nemo-like Kinase in Liver Cancer." *Cancer Letters* **344** (2): 204–11.

Sheng, Yanrui, Jianbo Li, Chengcheng Zou et al. 2014. "Downregulation of miR-101-3p by Hepatitis B Virus Promotes Proliferation and Migration of Hepatocellular Carcinoma Cells by Targeting Rab5a." *Archives of Virology* **159** (9): 2397–410.

Shi, Cailing, and Xudong Xu. 2013. "MicroRNA-22 Is down-Regulated in Hepatitis B Virus-Related Hepatocellular Carcinoma." *Biomedicine & Pharmacotherapy- Biomedecine & Pharmacotherapie* **67** (5): 375–80.

Shih, Tsung-Chieh, Yin-Jing Tien, Chih-Jen Wen et al. 2012. "MicroRNA-214 Downregulation Contributes to Tumor Angiogenesis by Inducing Secretion of the Hepatoma-Derived Growth Factor in Human Hepatoma." *Journal of Hepatology* **57** (3): 584–91.

Shimizu, Satoshi, Tetsuo Takehara, Hayato Hikita et al. 2010. "The Let-7 Family of microR-NAs Inhibits Bcl-xL Expression and Potentiates Sorafenib-Induced Apoptosis in Human Hepatocellular Carcinoma." *Journal of Hepatology* **52** (5): 698–704.

Shrivastava, Shubham, Anupam Mukherjee, and Ratna B Ray. 2013. "Hepatitis C Virus Infection, microRNA and Liver Disease Progression." *World Journal of Hepatology* **5** (9): 479–86.

Sidharthan, Sreetha, and Shyam Kottilil. 2014. "Mechanisms of Alcohol-Induced Hepatocellular Carcinoma." *Hepatology International* **8** (2): 452–7.

Stroffolini, T. 2005. "Etiological Factor of Hepatocellular Carcinoma in Italy." *Minerva Gastroenterologica E Dietologica* **51** (1): 1–5.

Su, Chenhe, Zhaohua Hou, Cai Zhang et al. 2011. "Ectopic Expression of microRNA-155 Enhances Innate Antiviral Immunity against HBV Infection in Human Hepatoma Cells." *Virology Journal* **8** (July): 354.

Su, Hang, Jian-Rong Yang, Teng Xu et al. 2009. "MicroRNA-101, down-Regulated in Hepatocellular Carcinoma, Promotes Apoptosis and Suppresses Tumorigenicity." *Cancer Research* **69** (3): 1135–42.

Suzuki, Kunio, Norio Hayashi, Yukinori Yamada et al. 1994. "Expression of the c-Met Protooncogene in Human Hepatocellular Carcinoma." *Hepatology (Baltimore, Md.)* **20** (5): 1231–36.

Sylvestre, Yannick, Vincent De Guire, Emmanuelle Querido et al. 2007. "An E2F/miR-20a Autoregulatory Feedback Loop." *The Journal of Biological Chemistry* **282** (4): 2135–43.

Tan, Weiqi, Yang Li, Seng-Gee Lim et al. 2014. "miR-106b-25/miR-17-92 Clusters: Polycistrons with Oncogenic Roles in Hepatocellular Carcinoma." *World Journal of Gastroenterology* **20** (20): 5962–72.

Tannapfel, Andrea, Dorothee Grund, Alexander Katalinic et al. 2000. "Decreased Expression of p27 Protein Is Associated with Advanced Tumor Stage in Hepatocellular Carcinoma." *International Journal of Cancer* **89** (4): 350–5.

Testino, Gianni, Silvia Leone, and Paolo Borro. 2014. "Alcohol and Hepatocellular Carcinoma: A Review and a Point of View." *World Journal of Gastroenterology: WJG* **20** (43): 15943–54.

Thibault, Patricia A., Adam Huys, Yalena Amador-Cañizares et al. 2015. "Regulation of Hepatitis C Virus Genome Replication by Xrn1 and MicroRNA-122 Binding to Individual Sites in the 5' Untranslated Region." *Journal of Virology* **89** (12): 6294–311.

Tian, Qi, Linhui Liang, Jie Ding et al. 2012. "MicroRNA-550a Acts as a pro-Metastatic Gene and Directly Targets Cytoplasmic Polyadenylation Element-Binding Protein 4 in Hepatocellular Carcinoma." *PloS One* **7** (11): e48958.

Tian, Yi, Weibing Yang, Jianxun Song et al. 2013. "Hepatitis B Virus X Protein-Induced Aberrant Epigenetic Modifications Contributing to Human Hepatocellular Carcinoma Pathogenesis." *Molecular and Cellular Biology* **33** (15): 2810–16.

Tomimaru, Y., H. Eguchi, H. Nagano, H. et al. 2010. "MicroRNA-21 Induces Resistance to the Anti-Tumour Effect of Interferon-α/5-Fluorouracil in Hepatocellular Carcinoma Cells." *British Journal of Cancer* **103** (10): 1617–26.

Tsai, Wei-Chih, Paul Wei-Che Hsu, Tsung-Ching Lai et al. 2009. "MicroRNA-122, a Tumor Suppressor microRNA That Regulates Intrahepatic Metastasis of Hepatocellular Carcinoma." *Hepatology (Baltimore, Md.)* **49** (5): 1571–82.

Tsang, Wing Pui, and Tim Tak Kwok. 2008. "Let-7a microRNA Suppresses Therapeutics-Induced Cancer Cell Death by Targeting Caspase-3." *Apoptosis: An International Journal on Programmed Cell Death* **13** (10): 1215–22.

Wang, Bo, Shu-Hao Hsu, Wendy Frankel et al. 2012. "Stat3-Mediated Activation of microRNA-23a Suppresses Gluconeogenesis in Hepatocellular Carcinoma by down-Regulating Glucose-6-Phosphatase and Peroxisome Proliferator-Activated Receptor Gamma, Coactivator 1 Alpha." *Hepatology (Baltimore, Md.)* **56** (1): 186–97.

Wang, Bo, Shu-Hao Hsu, Sarmila Majumder et al. 2010. "TGFβ Mediated Upregulation of Hepatic miR-181b Promotes Hepatocarcinogenesis by Targeting TIMP3." *Oncogene* **29** (12): 1787–97.

Wang, Chun-Mei, Yan Wang, Chun-Guang Fan et al. 2011. "miR-29c Targets TNFAIP3, Inhibits Cell Proliferation and Induces Apoptosis in Hepatitis B Virus-Related Hepatocellular Carcinoma." *Biochemical and Biophysical Research Communications* **411** (3): 586–92.

Wang, Jian, Jingwu Li, Junling Shen et al. 2012. "MicroRNA-182 Downregulates Metastasis Suppressor 1 and Contributes to Metastasis of Hepatocellular Carcinoma." *BMC Cancer* **12**: 227.

Wang, Lei, Xiang Zhang, Lin-Tao Jia et al. 2014. "C-Myc-Mediated Epigenetic Silencing of MicroRNA-101 Contributes to Dysregulation of Multiple Pathways in Hepatocellular Carcinoma." *Hepatology (Baltimore, Md.)* **59** (5): 1850–63.

Wang, Ruizhi, Na Zhao, Siwen Li et al. 2013. "MicroRNA-195 Suppresses Angiogenesis and Metastasis of Hepatocellular Carcinoma by Inhibiting the Expression of VEGF, VAV2, and CDC42." *Hepatology (Baltimore, Md.)* **58** (2): 642–53.

Wang, Saifeng, Lipeng Qiu, Xiaoli Yan et al. 2012. "Loss of microRNA 122 Expression in Patients with Hepatitis B Enhances Hepatitis B Virus Replication through Cyclin G(1)-Modulated P53 Activity." *Hepatology (Baltimore, Md.)* **55** (3): 730–41.

Wang, Wen, Lan-Juan Zhao, Ye-Xiong Tan et al. 2012. "MiR-138 Induces Cell Cycle Arrest by Targeting Cyclin D3 in Hepatocellular Carcinoma." *Carcinogenesis* **33** (5): 1113–20.

Wang, Wen-Tao, and Yue-Qin Chen. 2014. "Circulating miRNAs in Cancer: From Detection to Therapy." *Journal of Hematology & Oncology* **7**: 86.

Wang, Yanling, Li Jiang, Xiong Ji et al. 2013. "Hepatitis B Viral RNA Directly Mediates down-Regulation of the Tumor Suppressor microRNA miR-15a/miR-16-1 in Hepatocytes." *The Journal of Biological Chemistry* **288** (25): 18484–93.

Wang, Yu, Alvin T. C. Lee, Joel Z. I. Ma et al. 2008. "Profiling microRNA Expression in Hepatocellular Carcinoma Reveals microRNA-224 up-Regulation and Apoptosis Inhibitor-5 as a microRNA-224-Specific Target." *The Journal of Biological Chemistry* **283** (19): 13205–15.

Wang, Yu, Yiwei Lu, Soo Ting Toh et al. 2010. "Lethal-7 Is down-Regulated by the Hepatitis B Virus X Protein and Targets Signal Transducer and Activator of Transcription 3." *Journal of Hepatology* **53** (1): 57–66.

Wang, Zifeng, Sheng Lin, Julia Jun Li et al. 2011. "MYC Protein Inhibits Transcription of the microRNA Cluster MC-Let-7a-1~let-7d via Noncanonical E-Box." *The Journal of Biological Chemistry* **286** (46): 39703–14.

Wei, Xufu, Tingxiu Xiang, Guosheng Ren et al. 2013. "miR-101 Is down-Regulated by the Hepatitis B Virus X Protein and Induces Aberrant DNA Methylation by Targeting DNA Methyltransferase 3A." *Cellular Signaling* **25** (2): 439–46.

Wong, Carmen Chak-Lui, Chun-Ming Wong et al. 2011. "The microRNA miR-139 Suppresses Metastasis and Progression of Hepatocellular Carcinoma by down-Regulating Rho-Kinase 2." *Gastroenterology* **140** (1): 322–31.

Wong, Queenie W.-L., Arthur K.-K. Ching, Anthony W.-H. Chan et al. 2010. "MiR-222 Overexpression Confers Cell Migratory Advantages in Hepatocellular Carcinoma through Enhancing AKT Signaling." *Clinical Cancer Research: An Official Journal of the American Association for Cancer Research* **16** (3): 867–75.

Wu, G., F. Yu, Z. Xiao et al. 2011. "Hepatitis B Virus X Protein Downregulates Expression of the miR-16 Family in Malignant Hepatocytes in Vitro." *British Journal of Cancer* **105** (1): 146–53.

Wu, Hui Chen, and Regina Santella. 2012. "The Role of Aflatoxins in Hepatocellular Carcinoma." *Hepatitis Monthly* **12** (10 HCC): e7238.

Xie, Qing, Xiangmei Chen, Fengmin Lu et al. 2012. "Aberrant Expression of microRNA 155 May Accelerate Cell Proliferation by Targeting Sex-Determining Region Y Box 6 in Hepatocellular Carcinoma." *Cancer* **118** (9): 2431–42.

Xiong, Yujuan, Jian-Hong Fang, Jing-Ping Yun et al. 2010. "Effects of microRNA-29 on Apoptosis, Tumorigenicity, and Prognosis of Hepatocellular Carcinoma." *Hepatology (Baltimore, Md.)* **51** (3): 836–45.

Xu, Teng, Ying Zhu, Yujuan Xiong et al. 2009. "MicroRNA-195 Suppresses Tumorigenicity and Regulates G1/S Transition of Human Hepatocellular Carcinoma Cells." *Hepatology (Baltimore, Md.)* **50** (1): 113–21.

Xu, Xiaojie, Zhongyi Fan, Lei Kang et al. 2013. "Hepatitis B Virus X Protein Represses miRNA-148a to Enhance Tumorigenesis." *The Journal of Clinical Investigation* **123** (2): 630–45.

Xu, Yonghua, Yong An, Yun Wang et al. 2013. "miR-101 Inhibits Autophagy and Enhances Cisplatin-Induced Apoptosis in Hepatocellular Carcinoma Cells." *Oncology Reports* **29** (5): 2019–24.

Yang, Fu, Yixuan Yin, Fang Wang et al. 2010. "miR-17-5p Promotes Migration of Human Hepatocellular Carcinoma Cells through the p38 Mitogen-Activated Protein Kinase-Heat Shock Protein 27 Pathway." *Hepatology (Baltimore, Md.)* **51** (5): 1614–23.

Yang, Pengyuan, Qi-Jing Li, Yuxiong Feng et al. 2012. "TGF-β-miR-34a-CCL22 Signaling-Induced Treg Cell Recruitment Promotes Venous Metastases of HBV-Positive Hepatocellular Carcinoma." *Cancer Cell* **22** (3): 291–303.

Yang, Wei, Ting Sun, Jianping Cao et al. 2012. "Downregulation of miR-210 Expression Inhibits Proliferation, Induces Apoptosis and Enhances Radiosensitivity in Hypoxic Human Hepatoma Cells in Vitro." *Experimental Cell Research* **318** (8): 944–54.

Yang, Xin, Lei Liang, Xiao-Fei Zhang et al. 2013. "MicroRNA-26a Suppresses Tumor Growth and Metastasis of Human Hepatocellular Carcinoma by Targeting Interleukin-6-Stat3 Pathway." *Hepatology (Baltimore, Md.)* **58** (1): 158–70.

Yang, Xin, Xiao-Fei Zhang, Xu Lu et al. 2014. "MicroRNA-26a Suppresses Angiogenesis in Human Hepatocellular Carcinoma by Targeting Hepatocyte Growth Factor-cMet Pathway." *Hepatology (Baltimore, Md.)* **59** (5): 1874–85.

Yang, Xue-Wei, Long-Juan Zhang, Xiao-Hui Huang et al. 2014. "miR-145 Suppresses Cell Invasion in Hepatocellular Carcinoma Cells: miR-145 Targets ADAM17." *Hepatology Research: The Official Journal of the Japan Society of Hepatology* **44** (5): 551–9.

Yang, Yang, Menghe Li, Su 'e Chang et al. 2014. "MicroRNA-195 Acts as a Tumor Suppressor by Directly Targeting Wnt3a in HepG2 Hepatocellular Carcinoma Cells." *Molecular Medicine Reports* **10** (5): 2643–8.

Yao, Jian, Linhui Liang, Shenglin Huang et al. 2010. "MicroRNA-30d Promotes Tumor Invasion and Metastasis by Targeting Galphai2 in Hepatocellular Carcinoma." *Hepatology (Baltimore, Md.)* **51** (3): 846–56.

Ying, Qiao, Linhui Liang, Weijie Guo et al. 2011. "Hypoxia-Inducible microRNA-210 Augments the Metastatic Potential of Tumor Cells by Targeting Vacuole Membrane Protein 1 in Hepatocellular Carcinoma." *Hepatology (Baltimore, Md.)* **54** (6): 2064–75.

Yuan, Qinggong, Komal Loya, Bhavna Rani et al. 2013. "MicroRNA-221 Overexpression Accelerates Hepatocyte Proliferation during Liver Regeneration." *Hepatology (Baltimore, Md.)* **57** (1): 299–310.

Yuen, Man-Fung, Jin-Lin Hou, Anuchit Chutaputti, and Asia Pacific Working Party on Prevention of Hepatocellular Carcinoma. 2009. "Hepatocellular Carcinoma in the Asia Pacific Region." *Journal of Gastroenterology and Hepatology* **24** (3): 346–53.

Zeng, Lingyao, Jian Yu, Tao Huang et al. 2012. "Differential Combinatorial Regulatory Network Analysis Related to Venous Metastasis of Hepatocellular Carcinoma." *BMC Genomics* **13** (Suppl 8): S14.

Zhang, Guang-ling, Yi-xuan Li, Shu-qi Zheng et al. 2010. "Suppression of Hepatitis B Virus Replication by microRNA-199a-3p and microRNA-210." *Antiviral Research* **88** (2): 169–75.

Zhang, Hongyi, Zhiqiang Feng, Rui Huang et al. 2014. "MicroRNA-449 Suppresses Proliferation of Hepatoma Cell Lines through Blockade Lipid Metabolic Pathway Related to SIRT1." *International Journal of Oncology* **45** (5): 2143–52.

Zhang, J., Y. Yang, T. Yang et al. 2010. "microRNA-22, Downregulated in Hepatocellular Carcinoma and Correlated with Prognosis, Suppresses Cell Proliferation and Tumourigenicity." *British Journal of Cancer* **103** (8): 1215–20.

Zhang, J.-P., C. Zeng, L. Xu et al. 2014. "MicroRNA-148a Suppresses the Epithelial-Mesenchymal Transition and Metastasis of Hepatoma Cells by Targeting Met/Snail Signaling." *Oncogene* **33** (31): 4069–76.

Zhang, Weiying, Guangyao Kong, Junping Zhang et al. 2012. "MicroRNA-520b Inhibits Growth of Hepatoma Cells by Targeting MEKK2 and Cyclin D1." *PloS One* **7** (2): e31450.

Zhang, Xiao, Shijie Hu, Xiang Zhang et al. 2014. "MicroRNA-7 Arrests Cell Cycle in G1 Phase by Directly Targeting CCNE1 in Human Hepatocellular Carcinoma Cells." *Biochemical and Biophysical Research Communications* **443** (3): 1078–84.

Zhang, Xiaoying, Shanrong Liu, Tingsong Hu et al. 2009. "Up-Regulated microRNA-143 Transcribed by Nuclear Factor Kappa B Enhances Hepatocarcinoma Metastasis by Repressing Fibronectin Expression." *Hepatology (Baltimore, Md.)* **50** (2): 490–9.

Zhang, Xiaoyong, Ejuan Zhang, Zhiyong Ma et al. 2011. "Modulation of Hepatitis B Virus Replication and Hepatocyte Differentiation by MicroRNA-1." *Hepatology (Baltimore, Md.)* **53** (5): 1476–85.

Zhang, Yanqiong, Xiaodong Guo, Lu Xiong et al. 2012. "MicroRNA-101 Suppresses SOX9-Dependent Tumorigenicity and Promotes Favorable Prognosis of Human Hepatocellular Carcinoma." *FEBS Letters* **586** (24): 4362–70.

Zhang, Yiliang, Wei Wei, Na Cheng et al. 2012. "Hepatitis C Virus-Induced up-Regulation of microRNA-155 Promotes Hepatocarcinogenesis by Activating Wnt Signaling." *Hepatology (Baltimore, Md.)* **56** (5): 1631–40.

Zhang, Yizhou, Shoichi Takahashi, Akiko Tasaka et al. 2013. "Involvement of microRNA-224 in Cell Proliferation, Migration, Invasion, and Anti-Apoptosis in Hepatocellular Carcinoma." *Journal of Gastroenterology and Hepatology* **28** (3): 565–75.

Zhao, Shuhua, Yuanyuan Zhang, Xiuxiu Zheng et al. 2015. "Loss of MicroRNA-101 Promotes Epithelial to Mesenchymal Transition in Hepatocytes." *Journal of Cellular Physiology* **230** (11): 2706–17.

Zheng, Changli, Jingjing Li, Qi Wang et al. 2015. "MicroRNA-195 Functions as a Tumor Suppressor by Inhibiting CBX4 in Hepatocellular Carcinoma." *Oncology Reports* **33** (3): 1115–22.

Zheng, Fang, Yi-Ji Liao, Mu-Yan Cai et al. 2012. "The Putative Tumour Suppressor microRNA-124 Modulates Hepatocellular Carcinoma Cell Aggressiveness by Repressing ROCK2 and EZH2." *Gut* **61** (2): 278–89.

Zhu, Qin, Zhiming Wang, Yu Hu et al. 2012. "miR-21 Promotes Migration and Invasion by the miR-21-PDCD4-AP-1 Feedback Loop in Human Hepatocellular Carcinoma." *Oncology Reports* **27** (5): 1660–8.

Zhu, Shuomin, Hailong Wu, Fangting Wu et al. 2008. "MicroRNA-21 Targets Tumor Suppressor Genes in Invasion and Metastasis." *Cell Research* **18** (3): 350–9.

Zhu, X.-M., L.-J. Wu, J. Xu et al. 2011. "Let-7c microRNA Expression and Clinical Significance in Hepatocellular Carcinoma." *The Journal of International Medical Research* **39** (6): 2323–9.

9 MicroRNA Regulation of Invasive Phenotype of Glioblastoma

Omkar Vinchure and Ritu Kulshreshtha

CONTENTS

Introduction

Glioblastoma Multiforme (GBM) is the most common, highly aggressive and lethal cancer of the central nervous system. Based on the World Health Organization (WHO) classification, it belongs to the grade IV astrocytoma and develops tumors in the brain and rarely in the spinal cord (Kuo et al., 2015; Louis et al., 2016). Further, a new classification was devised to improve our understanding of GBM with the potential of development of precise therapy. It defined GBM into various subtypes (classical, mesenchymal, proneural and neural), depending on the expression/ mutation analyses of specific genes such as EGFR, TP53, IDH1, PDGFR, PTEN and NF1 (Verhaak et al., 2010).

A high proliferation index and extensive invasion are the hallmarks of GBM leading to poor prognosis with the median overall survival being a mere approximate of 14 months after diagnosis (Adamson et al., 2010; Anton et al., 2012). The treatment modalities include surgical resection, followed by radiotherapy and chemotherapy, but recurrence almost always occurs in all patients leading to death. The therapy failure leading to decreased survival has been mainly attributed to incomplete surgical removal of the tumor (owing to location and diffuse boundaries of the tumor due to infiltration) (Brown et al., 2016), extreme tumor heterogeneity (both phenotypic and genotypic) (Eder et al., 2014), presence of cancer stem cells (CSCs) (contributing to tumor regeneration even after therapy) (Sundar et al., 2014) and existence of hypoxic regions (contributing to tumor aggressiveness and therapy resistance) (Yang et al., 2012). Because of the pathetic outcome of the current standard therapies for GBM treatment, novel approaches are being tested to treat this tumor such as suicide gene therapy, potentiation of immune system, oncolytic viruses and more recently microRNA based therapy (Hulou et al., 2016).

MicroRNAs (miRNAs) represent a class of short, non-coding RNA molecules that regulate the gene expression at the post-transcriptional level via physical interaction with their cognate mRNAs known as targets. Largely, based on the extent of complementarity with the target site (usually present in the 3'UTR or less frequently within the coding region or 5'UTR), miRNAs regulate gene expression either by promoting the degradation of target transcripts or by inhibiting their translation, hence, ultimately affecting the cellular phenotype (Bartel, 2009). Recent studies have shown that miRNAs may induce the expression of target genes as well, via interplay with RNA binding proteins, or direct interaction with elements in the promoter or 5'UTR to promote mRNA stability and translational activation (Vasudevan et al., 2007; Orang et al., 2014).

The miRNAs are now known to play an imperative role in cancer pathogenesis via regulation of various cancer hallmarks, cancer cell stemness, therapy resistance and immune modulation (Hayes et al., 2014). Depending on whether the miRNAs have a net pro-cancerous or a tumor-suppressive role, based on the set of mRNAs they regulate, miRNAs are classified as oncomiRs and tumor suppressors (Svoronos et al., 2016). Since the miRNAs are known to be aberrantly expressed in various cancer types, and play an active role in cancer pathogenesis, they are considered as promising agents for cancer diagnosis and therapy. The involvement of miRNAs in regulation of genes that mediate GBM invasion is beginning to be appreciated. This chapter presents an overview of the interplay between key pathways, various molecular factors and miRNAs that show concerted regulation of GBM invasiveness. Novel therapeutic miRNA and drug delivery approaches to treat GBM that are currently under development are also discussed.

9.1 MODES OF INVASION AND METASTASIS IN GBM

Like many other malignancies, GBM cells also invade healthy tissues, thus contributing to the pathogenesis. However, systemic metastasis, unlike other cancer types, is rare due to the hindrance posed by the blood-brain barrier and by the absence of lymphatic vessels. Intracranial metastases (usually within 2cm of the original tumor), leading to the development of multifocal and multicentric GBM, however, remain the major cause of tumor recurrence and fatality caused by GBM (Patil et al., 2012).

GBM cells usually migrate through the perivascular space (present around blood vessels), space between glial cells and neurons in brain parenchyma and rarely through cerebrospinal fluid (in case of spinal metastasis) (Cuddapah et al., 2014; Shah et al., 2010). GBM cells invade the surrounding healthy brain stroma by actuating a process named as Epithelial-to-Mesenchymal Transition (EMT), where cells lose their epithelial characteristics and acquire mesenchymal properties through a series of biochemical changes to attain a pro-invasive and cancer stem cell (CSC) phenotype, which is capable of both tumor initiation and maintenance and systemic metastasis. The EMT phenotype has also been associated with development of drug resistance, leading to tumor recurrence and poor prognosis (Lamouille et al., 2014). Another process, known as the mesenchymal-to-epithelial transition (MET), is responsible for enabling the invasion of cancer cells into the tissue stroma, thus aiding in colonization in different regions within brain that may be vital for patient's survival (Lee et al., 2014).

Overall, cancer cell infiltration of healthy tissue is a very complex and, yet, a tightly regulated process. A wide array of genes, complex protein interaction pathways and different classes of regulatory molecules have been reported to work in unison in response to signals promoting cell invasion and metastases.

9.2 EFFECTORS OF EPITHELIAL-TO-MESENCHYMAL TRANSITION AND INVASION IN GBM

9.2.1 CADHERINS AND EMT

Several sequential molecular changes are hallmarks of EMT. The process of EMT begins with the loss of cell-cell contact via disruption of desmosomes, gap junctions, adherens junctions and tight junctions (Lamouille et al., 2014). This is followed by the degradation of epithelial markers like E-Cadherin with a concomitant accumulation of mesenchymal markers like N-Cadherin, Vimentin and fibronectin (Lee et al., 2014). Low levels of E-Cadherin, while high levels of N-cadherin and Vimentin, have been associated with GBM progression and poor prognosis (Mariotti et al., 2007; Canel et al., 2013; Satelli et al., 2011). The downregulation of T-Cadherin (Cadherin 13), has also been implied in the reduction of epithelial characteristics during EMT to promote GBM cell invasion (Andreeva and Kutuzov; 2010). N-Cadherin, also known as Neural Cadherin, is a cell adhesion protein that promotes stabilization of tumor-associated blood vessels and its overexpression is associated with increased invasive ability of glioma (Mariotti et al., 2007). Vimentin, a mesenchymal marker and a regulator of wound healing and migration, is a member of the intermediate filament family that provides structural integrity. An important player known as Beta-catenin (β-catenin), which is a transcription factor and a downstream effector of the Wnt signaling and PI3K signaling pathways, maintains the epithelial integrity of cells by complexing with E-Cadherin at cell membrane. Under the influence of Wnt pathway promoting signals, and the loss of E-Cadherin, β-catenin translocates into the nucleus to activate transcription of key genes involved in the promotion of EMT and invasion (Lee et al., 2014). Ultimately, cells lose their apical-basal polarity which is triggered by reorganization of actin cytoskeleton and formation of structures known as invadopodia that impart enhanced motility, wound healing and invasive phenotype to the cells allowing the dissemination of cells from tumor bulk (Moreno-Bueno et al., 2008).

9.2.2 EXTRACELLULAR MATRIX AND MMPS IN GBM INVASION

Extracellular Matrix (ECM) is a collective term for the group of proteins secreted out by cells into the surrounding medium. ECM facilitates many functions like cell-cell adhesion, communication, and mechanical support, and thus, poses as a barrier to GBM cell invasion. It is composed of many protein families like collagen, fibronectin, laminin, hyaluronic acid and proteoglycans. A hallmark of invasive GBM is the degradation of ECM, which is essential for the cells to invade the organ stroma (Lu et al., 2011). Degradation of the ECM is achieved via the many protein degrading enzyme families like matrix metalloproteinases (MMPs), serine proteinases, cysteine proteinases and aspartic proteinases (Lu et al., 2011). MMPs, however, are the principle enzymes enabling almost complete ECM degradation. Notably, MMP-2 and MMP-9 contribute to promote cell invasion in GBM. The expression levels of these have been shown to increase significantly in invasive GBM as compared to normal healthy brain tissue and have been associated with poor prognosis in GBM patients (Nakada et al., 2003). Interestingly, knockdown of MMP-9 and cathepsin B in a glioma cell line brought about a decrease in pre-established tumor growth, invasion and angiogenesis (Lakka et al., 2004).

9.2.3 TRANSCRIPTIONAL CONTROL OF EMT IN GBM

The induction of EMT is majorly mediated by three families of EMT-activating Transcription factors (EMT-ATFs), namely ZEB (Zinc-finger enhancer binding), SNAI1 (Snail Family Transcriptional Repressor 1) and TWIST (Twist-related protein 1), the levels of which are shown to be highly expressed in invasive GBM and correlate with increased EMT, glioma grade and poor prognosis (Elias et al., 2005; Siebzehnrubl et al., 2013; Myung et al., 2014). They have been shown to activate the transcription of mesenchymal markers and inhibit the expression of the genes responsible for epithelial phenotype. While ZEB and TWIST family of TFs repress the transcription of E-cadherin by binding to E-box elements present in its promoter, the SNAIL family recruits chromatin-modifying machinery comprising of DNA Methyltransferases (DNMTs) to E-Cadherin promoter region to suppress the expression of E-Cadherin (Sanchez et al., 2012). These TFs also bring about induction in the expression of N-cadherin, Vimentin and MMPs to promote invasion (Sánchez-Tilló et al., 2012). Overall, these TFs play a key role in the EMT process, and thus, inhibition of these members has been considered as a strategy to inhibit GBM cell invasion.

9.3 ROLE OF miRNAs IN INVASION

Invasion in GBM is a multistep process that involves the participation of several molecular effectors that work in unison along discrete pathways. Many factors are known to govern invasion by recruiting transcriptional machinery to drive the expression of key genes and by modulating the cell microenvironment to promote cellular mobility. miRNAs too are known to intervene in the process of GBM invasion by affecting the EMT process via direct regulation of its effectors, or indirectly via regulation of key effectors of pathways that are active in GBM invasion (Table 9.1). Notably, some oncomiRs tend to show higher expression during GBM invasion, while tumor suppressor miRNAs are downregulated. Although distinct patterns of miRNA expression in the context of GBM invasiveness have emerged, the tumor heterogeneity, changing microenvironment and case-to-case variation has revealed varying miRNA signatures and unexpected outcomes. Thus, overall understanding of miRNA profile in the invasive setting would be essential for development of precision therapy.

9.3.1 MIRNA MEDIATED REGULATION OF EFFECTORS OF GBM INVASION

9.3.1.1 EMT TFs and Cadherins

Specific miRNAs have been reported to alter the expression of epithelial and mesenchymal markers such as E-Cadherin, N-Cadherin, Vimentin and EMT-TFs to mediate GBM cell invasion. miR-92a-3p, a miRNA overexpressed in glioma patients, was shown to directly target E-Cadherin. Its inhibition was shown to inhibit glioma cell migration and invasion (Song et al., 2016a). Similarly, miR-130b directly targeted PPAR-γ, decreased E-cadherin and increased the β-catenin levels thereby promoting the proliferation and invasion of glioma cells *in vitro* (Gu et al., 2016). The anti-invasive miRNAs such as let-7b/let-7i, miR-16, miR-517c, miR-663 and miR-154, and inhibition of pro-invasive miR-23b (targeting PYK2) inhibited migration and invasion in glioma cells by bringing about upregulation of E-cadherin and downregulation of pro-EMT genes such as N-cadherin, Vimentin, β-catenin and EMT-TFs (Slug and Snail) (Tian et al., 2015; Wang et al., 2014; Lu et al., 2015; Li et al., 2016; Zhao et al., 2016; Chen et al., 2014). Among these, let-7b, miR-16, miR-517c and miR-663 have been correlated to better prognosis of GBM patients.

Interestingly, some pharmacological inhibitors of invasion involved upregulation of miRNA candidates that negatively regulate EMT. For example, Erismodegib is a molecular inhibitor of SMO protein whose aberrant activating mutation leads to upregulation of the Sonic Hedgehog signaling pathway. It was shown to inhibit EMT in Glioblastoma Initiating Cells (GICs) by upregulating E-Cadherin and concomitantly downregulating N-Cadherin and EMT-TFs-(Snail, Slug and Zeb-1) via the induction of miR-200 family leading to inhibition of invasion *in vitro* (Fu et al., 2013). Likewise, AC1MMYR2, an inhibitor of the DICER mediated processing of pre-miR-21 into mature miR-21 (having a negative prognosis in GBM), was reported to inhibit GBM cell migration and invasion abilities marked by an increase in E-Cadherin and a decrease in N-Cadherin, β-catenin, ZEB1/2, and MMP-9 levels (Shi et al., 2013). Novel pharmacological intervention strategies, thus, may be developed to reduce EMT progression in GBM by modulation of specific miRNA levels.

TABLE 9.1

List of Invasion-Related miRNAs, Their Targets and Prognostic Significance in Glioblastoma

MicroRNA	Effect on Invasion	Target	Other Effects	Prognosis	Cell Lines, Animal Models	Citation
let-7b	Inhibition	E2F2	Upregulation of E-Cadherin	Positive	U87, U251	Tian et al., 2015; Song et al., 2016b; Mao et al.
		IKBKE				
let-7i	Inhibition		Upregulation of E-Cadherin	–	U251, U287	Tian et al., 2016; Yan et al., 2015
miR-125b	Promotion	PIAS3	Upregulation of MMP-2, MMP-9	Positive	GSCs	Shi et al., 2014; Henriksen et al., 2014
miR-125b-1	Inhibition			–	A172	Gu et al., 2014
miR-130b	Promotion	PPAR γ		–	U251, U87, SNB19, LN229	Gu et al., 2016
miR-137	Inhibition	RTVP-1	Downregulation of MMP, CXCR4, SHH ligand	Positive	GSC	Bier et al., 2013; Li et al., 2016
miR-144	Promotion			Positive	A172	Gu et al., 2014
miR-144-3p	Inhibition	MET		Positive	U87, LN229, U251, LN18, H4, nude mice (flanks)	Lan et al., 2015
miR-154	Inhibition	Wnt5a	Upregulation of E-Cadherin, downregulation of N-Cadherin, Vimentin, SNAI1	–	U87, U251, A172	Zhao et al., 2016
miR-16	Inhibition		Downregulation of Vimentin, E-Cadherin, Slug	Positive	U87, U251	Wang et al., 2014; Ye et al., 2017
miR-17-5p	Promotion	PTEN	Upregulation of VEGF, HIF1α	–	U87, U343	Li et al., 2012
miR-181d	Inhibition	MGMT	Downregulation of Beta catenin	Positive	LN229, SNB19, U87, A172, LN308, nude mice (flank)	Shi et al., 2014; Zhang et al., 2012
miR-182	Inhibition	HIF2A, BCL2L12		Positive	U87, female SCID mice (intracranial)	Kouri et al., 2015

(Continued)

TABLE 9.1 (CONTINUED)
List of Invasion-Related miRNAs, Their Targets and Prognostic Significance in Glioblastoma

MicroRNA	Effect on Invasion	Target	Other Effects	Prognosis	Cell Lines, Animal Models	Citation
miR-200a, b, c	Inhibition		Upregulation of E-Cadherin, downregulation of N-Cadherin, Snail, Slug, Zeb1, β-catenin	–	GIC, LN229, SNB19, U87, A172, LN308, H4, nude mice (flank)	Fu et al., 2013; Shi et al., 2014
miR-21	Promotion	RECK, TIMP3, VHL, PPARα	Downregulation of E-Cadherin, Upregulation of N-Cadherin, β-catenin, Zeb1, Zeb2, MMP-2, MMP-9, Vimentin	Negative	LN229, SNB19, U87, A172, LN308, U251, nude mice (mammary fat pads)	Shi et al., 2013; Gabriely et al., 2008; Dong et al., 2012; Ma et al., 2012; Zhang et al., 2014; Shi et al., 2014; Lakomy et al., 2011
miR-211	Inhibition		Downregulation of IR induced MMP-9	–	U87, nude mice (intracerebral)	Asuthkar et al., 2012
miR-218	Inhibition	LEF1		–	U251, U87, SNB19, LN229	Liu et al., 2012
miR-23b	Inhibition	PYK2	Upregulation of Zeb1, β-catenin; downregulation of E-Cadherin	–	A172, T98G, U87, SNB19, U251, SF767, G112MS, 293T, LN229, ex vivo orgnaotype invasion study, nude mice (intracranial)	Loftus et al., 2012; Chen et al., 2014; Shi et al., 2014
miR-302-367 cluster	Inhibition	CXCR4		–	TG1, TG6, GB1, #1056, male NOD.CB17-Prkdcscid/NCrHsd mice (striatum)	Fareh et al., 2012
miR-320a	Inhibition	IGF-1R	Upregulation of Akt	Positive	U87, LN229, U251, SHG44, H4, nude mice (brain)	Guo et al., 2014
miR-326	Inhibition	SMO		Positive	U87, SHG44, U251, T98G, HEK293T, nude mice (intracranial)	Du et al., 2015; Qiu et al., 2013

(Continued)

TABLE 9.1 (CONTINUED)
List of Invasion-Related miRNAs, Their Targets and Prognostic Significance in Glioblastoma

MicroRNA	Effect on Invasion	Target	Other Effects	Prognosis	Cell Lines, Animal Models	Citation
miR-328	Promotion	sFRP1		Positive	A172, T98G, TP365MG	Delic et al., 2014; Wu et al., 2012
miR-330	Promotion		Downregulation of SH3GL2	—	U87, GSC, nude mice (subcutaneous, orthotopic)	Yao et al., 2014
miR-330-5p	Inhibition	ITGA5		—	U87, U251, U373	Feng et al., 2017
miR-34a	Inhibition	Rictor	Promoting degradation of β-catenin	—	HNGC-2, NSG-K16, NOD-SCID mice (flank)	Rathod et al., 2014
miR-363	Promotion		Upregulation of MMP-2, MMP-9	—	Patient derived primary glioma cell lines	Conti et al., 2016
miR-373	Inhibition	MT1-MMP, CD44, TGFBR2	Downregulation of MMP-2	Negative	U87, U251	Lu et al., 2015; Wei et al., 2016; Li et al., 2014
miR-422a	Inhibition	PIK3CA		Positive	U373, TJ905, U251, SHG44, nude mice (flank)	Liang et al., 2016
miR-449a	Inhibition	MAZ	Downregulation of MMP-2, MMP-9	—	U87, U251, GSC, nude mice (flank)	Yao et al., 2015
miR-451	Inhibition		Downregulation of Akt, Upregulation of MMP-2, MMP-9	—	A172, LN229, U251, U87, SNB19	Nan et al., 2010, Zhao et al., 2017
miR-491-5p	Inhibition	MMP-9		—	U87, U251	Yan et al., 2011
miR-513c	Inhibition	LRP6		—	LN319, SNB19, LN18, U251MG, U118MG, T98G, U373MG, LN382T	Xu et al., 2015

(Continued)

TABLE 9.1 (CONTINUED)
List of Invasion-Related miRNAs, Their Targets and Prognostic Significance in Glioblastoma

MicroRNA	Effect on Invasion	Target	Other Effects	Prognosis	Cell Lines, Animal Models	Citation
miR-517c	Inhibition		Downregulation of N-Cadherin, Vimentin, Snail	Positive	U87, U251, HEK293T, nude mice (flank)	Lu et al., 2015
miR-520c	Inhibition	MT1-MMP	Downregulation of MMP-2	—	U87	Lu et al., 2015
miR-566	Promotion	VHL	Promoting degradation of Beta catenin, Upregulation of VEGF	—	U87, LN229, SNB19, LN308, U251, nude mice (intracranial)	Zhang et al., 2014; Xiao et al., 2016
miR-577	Inhibition	LRP6, β-catenin		—	U87, T98G, G112, U373MG, U251MG, LN18, nude mice (flank)	Zhang et al., 2016
miR-603	Promotion	CTNNBIP1	Promoting nuclear translocation of β-catenin	—	U8vIII, U87-MG, T98G, LN229, U138, U378, nude mice (flank)	Guo et al., 2015
miR-608	Inhibition	MIF		Positive	U87, U251, GSC	Wang et al., 2016
miR-663	Inhibition	TGFB1, CXCR4, PIK3CD	Downregulation of MMP-2, Upregulation of E-Cadherin	Positive	HEB, U87MG, Patient-derived primary cell line, male SCID mice	Li et al., 2016; Shi et al., 2015; Shi et al., 2014
miR-7	Inhibition	PI3K, FAK	Downregulation of MMP-2, MMP-9	—	U87, U251, CHG5, TJ899, TJ905, nude mice (subcutaneous, multi site)	Wu et al., 2011; Liu et al., 2014
miR-9	Inhibition	REB, MK2, MK3, SRF		Positive	T98G, SF295, U251	Ben-Hamo et al., 2016
miR-9	Context (density) dependent	PTCH1	Upregulation of E-Cadherin	—	9L, U87, T98G	Munoz et al., 2015; Katakowski et al., 2016
miR-92a-3p	Promotion	E-Cadherin		—	U87, U251, nude mice (flank)	Song et al., 2016a

9.3.1.2   Matrix Metalloproteinases

Since matrix metalloproteinases (MMPs) play a major role in the invasion process, their levels remain tightly regulated. Several miRNAs have been shown to affect the invasive phenotype by targeting specific MMP proteins in GBM miR-21, an oncomiR in GBM, was reported to promote GBM cell migration and invasion abilities by targeting RECK and TIMP3 genes, both of which are inhibitors of MMPs (Gabriely et al., 2008). In another study, miR-21 was reported to increase the expression of the mesenchymal marker, Vimentin, in GBM (Ma et al., 2012). Yan et al., 2011 identified MMP-9 specific miRNA signature in GBM and showed that miR-491-5p directly targets MMP-9 and inhibits invasion (Yan et al., 2011). miR-7 and miR-663 were also shown to inhibit GBM cell invasion by bringing about downregulation of MMP-2 or MMP-9 levels (Wu et al., 2011; Li et al., 2016). An interesting feedback loop was observed between miR-211 and MMP-9. miR-211 was shown to inhibit invasion in GBM stem cells via suppression of MMP-9 levels while MMP-9 epigenetically silenced miR-211 levels by promoting promoter methylation (Asuthkar et al., 2012). Similarly, inhibition of miR-363 or miR-125b miRNAs reduced invasiveness of primary GBM cell cultures or GSCs, respectively, by downregulating MMP-2 and MMP-9 levels (Conti et al., 2016; Shi et al., 2014). Paradoxically, miR-125b was reported to have a positive prognostic value in GBM patients with low nestin levels (Henriksen et al., 2014).

Besides the role of invasion effectors like MMP9 in epigenetic regulation of tumor suppressor miRNA (miR-211), RNA binding proteins have also been shown to regulate miRNA expression and thus invasion. Higher expression of an RNA binding protein LIN28A has been linked to poor prognosis in GBM patients (Qin et al., 2014) and its expression is GBM grade dependent (Mao et al., 2013). LIN28A was shown to bind to the let-7 family of miRNA precursors and inhibit the maturation and subsequent action of these miRNAs. LIN28A inhibited let-7b and let-7g expression, with a concomitant upregulation of HMGA2 and pro-EMT gene SNAI1 to promote invasion in GBM (Mao et al., 2013). Subsequently, a recent study showed that let-7b that has a prognostic significance in GBM, inhibits migration and invasiveness of GBM cell lines by targeting E2F2 (Song et al., 2016b).

An interesting study, however, points out to a cancer type-dependent role of miRNAs in regulating MMP expression. miR-373 (with a negative prognostic value in GBM) and miR-520c were shown to inhibit the expression of MMP2 in U87MG GBM cell line via directly targeting MT1-MMP (Lu et al., 2015). However, in contrast these miRNAs induced the expression of MMP-9 in HT1080 fibrosarcoma cell line to promote cell migration, although this effect was not observed in GBM cell line (Lu et al., 2015). Thus, miRNA mediation of cancer cell migration and invasion could be cancer-type specific, and it should be carefully assessed during design of therapeutic intervention strategies.

9.3.2   miRNA Mediated Regulation of Signaling Pathways in GBM Invasion

Many pathways are aberrantly regulated during cancer development and maintenance. The complex nature of some of these pathways, and their concerted mode of action via players (some of which are effectors of multiple pathways) and the accumulation of mutations in members of these pathways, together pose difficulties in devising a molecular intervention strategy in the treatment of cancer. The interactions of miRNAs with the pathways involved in GBM invasion are discussed below.

9.3.2.1 Wnt Signaling Pathway

Wnt signaling is an important pathway governing embryonic development, cell cycle regulation and maintenance of homeostasis. The Wnt signaling is activated by Wnt ligands that bind to their cognate receptor molecules to initiate three different signaling cascades—Canonical Wnt pathway (β-catenin dependent), Planar cell polarity pathway (mediated by Dishevelled via the action of RHOA, RAC and Cdc42), and Calcium flux dependent pathway (mediated by CaMK2, JNK and PKC) (Duchartre et al., 2016). In the absence of Wnt ligands, β-catenin gets phosphorylated as part of a multiprotein destruction complex containing Dishevelled, adenomatous polyposis coli (APC), Axin and GSK3β which leads to its destruction by the proteasome pathway. However, the interaction of Wnt with Fzd/LRP leads to the inactivation of the destruction complex. The unphosphorylated β-catenin then translocates to the nucleus to bring about transcriptional activation of genes involved in the Wnt signaling (Paw et al., 2015). The Wnt signaling regulates key functions of epithelial cells and its hyper activation has been associated with enhanced tumor growth, metastasis and chemo resistance in GBM (Tai et al., 2015; Yi et al., 2016). The deregulation of Wnt signaling pathway in brain cancers including GBM is, at least in part, an outcome of genetic changes in Wnt signaling components like APC, Axin, or β-catenin, E-cadherin or FAT1 cadherin (Nikuseva-Martić et al., 2007; Morris et al., 2013). Various soluble Frizzled-related proteins (sFRP1, sFRP2 and sFRP4) have been shown to inhibit GBM cell migration by bringing about phosphorylation of β-catenin or by of Wnt/Ca^{2+} pathway (antagonist of Wnt/β-catenin pathway) (Roth et al., 2000; Bhuvanalakshmi et al., 2015). Wnt signaling was also shown to positively affect GBM cell migration and invasion via the upregulation of MMP-2 (Kamino et al., 2011).

Inhibition of Wnt pathway is considered as a promising therapeutic option for GBM treatment. The use of miR-21 inhibitor was employed to reduce the canonical Wnt signaling in GBM by promotion of β-catenin degradation by von Hippel-Lindau (VHL) or by inhibiting its translocation into the nucleus, which subsequently inhibited the EGFR/AKT signaling (Zhang et al., 2014). miR-21 inhibition is especially important because of the existence of a positive feedback mechanism between miR-21 and β-catenin that enables constant expression of miR-21 to ultimately keep the EGFR signaling active which is generally observed in GBM (Han et al., 2012). Interestingly, miR-328 was identified as the main upregulated miRNA in infiltrating GBM cells *in vivo* and was shown to promote cell invasion by targeting sFRP1 especially in low-grade astrocytoma (Delic et al., 2014). Similarly, miR-603 activated the Wnt signaling pathway by promoting the nuclear translocation of β-catenin by targeting β-catenin interacting protein 1, CTNNBIP1 (Guo et al., 2015). In contrast, miR-34a was shown to inhibit glioma invasion by promoting the degradation of β-catenin mediated by activation of GSK-3β via the downregulation of p-AKT by Rictor (Rathod et al., 2014). Also, miR-577 and miR-513c inhibited the Wnt signaling by directly targeting low-density lipoprotein receptor-related protein 6 (LRP6) in glioma (Zhang et al., 2015; Xu et al., 2015). miR-200a and miR-577 were also shown to directly target β-catenin to mediate Wnt signaling suppression (Su et al., 2012; Zhang et al., 2015). Similarly, miR-181d that was associated with a better prognosis was shown to block the β-catenin/CBP transcriptional activity, and thus, invasion by directly targeting β-catenin and CREBBP in GBM (Shi et al., 2014; Zhang et al., 2012). Notably, miRNAs promoting Wnt signaling (miR-21, miR-328, miR-603) were shown to be upregulated while those inhibiting it (miR-34a, miR-577, miR-513c and miR-200b) were downregulated in GBM cell lines and patients.

Overall, it appears that β-catenin is the main target of the various miRNAs targeting the Wnt pathway. This was found to be at various levels—direct targeting of β-catenin (miR-181d, miR-200b, miR-577), affecting β-catenin degradation (miR-21, miR-34a) or affecting its nuclear translocation or transcriptional activity (miR-21, miR-181d, miR-603). Interestingly, BASI (also known as NSC141562), a small molecule inhibitor of cell proliferation and invasion, downregulated the expression of miR-21 and miR-23b, and upregulated miR-200a and miR-181d, which consequently led to the downregulation of β-catenin and inhibition of its activity in glioma cell lines and in U87MG subcutaneous glioma mouse model (Shi et al., 2014). Overall, miRNAs play a major role in the Wnt signaling pathway by affecting the levels/activity of the effectors to regulate GBM cell migration and invasion (see Figure 9.1).

9.3.2.2 PI3K/Akt Signaling Pathway

Phosphatidylinositol-3-kinase (PI3K) family comprises lipid kinases that are activated by receptors from Receptor Tyrosine Kinase (RTKs), GPCRs or oncogenes such as Ras (Li et al., 2016). After ligand binding, the RTKs phosphorylate the regulatory subunit of PI3K. PI3K mediated phosphorylation of PIP2 then leads to the formation of PIP3 that acts as a secondary messenger. This event is negatively regulated by PTEN. PIP3 then interacts with the PH domain of Akt, a protein kinase that shares homology with protein kinase A (PKA) and protein kinase C (PKC), to cause aggregation of Akt at the inner membrane. Finally, PDK1 phosphorylates Akt, thereby activating it to regulate downstream cellular effects like cell cycle, growth, differentiation, survival, angiogenesis, migration, etc. via a phosphorylation cascade (Li et al., 2016). Abnormalities in this pathway and its hyper-activation have been linked to many cancer types including GBM (Li et al., 2016). It has been the focus of many studies on glioma invasion wherein it was shown to induce the levels of pro-EMT and invasion transcription factors and effectors such as MMP2, MMP9 and β-catenin to promote invasion (Paw et al., 2015).

Several endogenous molecules regulate PI3K/Akt pathway. For instance, IMP2, a protein whose expression is upregulated in GBM, has been shown to activate PI3K/Akt pathway by upregulating IGF2, and to promote invasion in GBM by upregulation of N-Cadherin and Vimentin with a concomitant downregulation of E-Cadherin (Mu et al., 2015). It was also reported that a chemokine receptor CXCR4 activated PI3K/Akt pathway to upregulate the expression of Twist to mediate an increase in MMP-9 and N-Cadherin and a decrease in E-Cadherin resulting in increased EMT and invasion in GBM cells (Yao et al., 2016). Thus, PI3K/Akt pathway appears to be a crucial pathway that regulates GBM cell migration and invasion. Several pharmacological inhibitors such as Sulindac, LY294002, FTY720, and natural compounds

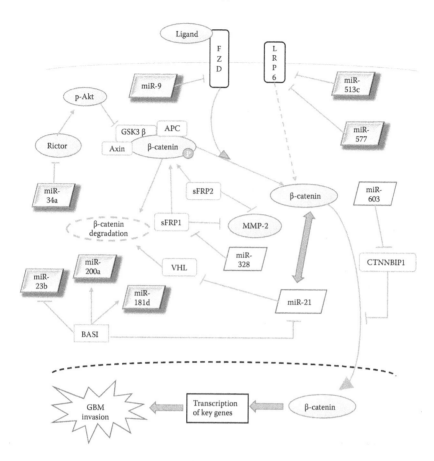

FIGURE 9.1 Interactions of Wnt/β-catenin signaling pathway components and miRNAs in the regulation of GBM invasion. Pro-invasive genes are shown in gray eclipses while anti-invasive genes are shown in rounded rectangles. The highlighted parallelograms represent miRNAs suppressing GBM invasion while plain parallelograms represent pro-invasive miRNAs. Bold and dotted connections represent direct and indirect effects, respectively.

like Resveratrol have shown their anti-invasive effects in GBM by bringing about inhibition of PI3K/Akt signaling (Jiao et al., 2015; Paw et al., 2015).

The role of miRNAs in regulation of PI3K/Akt pathway was recently shown in GBM (see Figure 9.2). For example, miR-330 activated the PI3K/Akt pathway by targeting SH3GL2 in GSCs to enhance cell invasion (Yao et al., 2014). Other miRNAs have been shown to inhibit the PI3K/Akt pathway by targeting PI3K itself (miR-7) or directly targeting the positive regulators of the pathway (miR449a targets Myc-associated Zinc-finger protein MAZ, an inducer of PI3K; miR-422a targets PIK3CA; miR-320a targets insulin-like growth factor-1 receptor (IGF1R) and miR-608 targets oncogenic migration inhibitory factor (MIF)) (Liu et al., 2014; Yao et al., 2015; Liang et al., 2016; Guo et al., 2014; Wang et al., 2016). Interestingly, high levels of some miRNAs (miR-320a, miR-422a and miR-608) involved in inhibition of PI3K/Akt

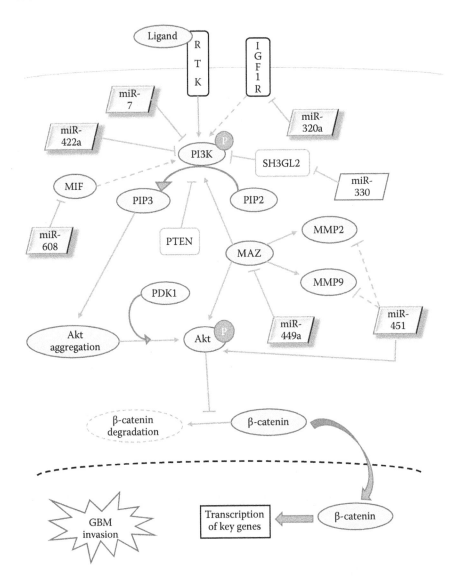

FIGURE 9.2 Interactions of PI3K/Akt signaling pathway components and miRNAs in the regulation of GBM invasion. Pro-invasive genes are shown in gray eclipses while anti-invasive genes are shown in rounded rectangles. The highlighted parallelograms represent miRNAs suppressing GBM invasion while plain parallelograms represent pro-invasive miRNAs. Bold and dotted connections represent direct and indirect effects, respectively.

signaling, and thus, invasion were shown to correlate with better prognosis in GBM patients (Table 9.1).

Xenograft experiments conducted in nude mice showed enhanced survival of mice overexpressing miR-449a, or miR-320a while miR-7 was shown to inhibit xenograft growth (Yao et al., 2015; Guo et al., 2014; Liu et al., 2014). Similarly, miR-422a was associated with a poor survival in GBM patients (Liang et al., 2016). Thus, miRNAs mediate the activity of PI3K/Akt pathway by direct interaction with its functional components or its regulators to affect GBM cell migration and invasion.

9.3.2.3 Hedgehog-GLI1 Signaling Pathway

This pathway is initiated by three distinct classes of ligands namely Sonic hedgehog (Shh), Indian hedgehog (Ihh) and Desert hedgehog (Dhh) that bind to their cognate transmembrane receptor molecules termed as Patched (PTCH) to abrogate its repressive effect on Smoothened (SMO). Free SMO then prevents the degradation of Glioma-Associated Oncogene Homolog 1 (GLI1), which then translocates into the nucleus and drives the transcription of key genes (Paw et al., 2015). The Hedgehog-GLI1 signaling is essential for tissue patterning, development, cell proliferation and stem cell maintenance (Gorojankina, 2016).

Increased activation of Hedgehog/GLI1 signaling pathway, involving alterations in its components have been implied in GBM. For instance, alternatively spliced transcript of GLI1, yields a truncated version (tGLI1) that exhibits a gain of function of GLI1 and leads to transcriptional regulation of genes not regulated by GLI1 and promotes aggressive cancer phenotypes by promoting motility, invasion and vascularization (Carpenter et al., 2012). In GBM, increased SHH ligand activity has been linked to the activation of Hedgehog signaling specifically in progenitor cells (Ehtesham et al., 2007). GLI1 has been reported to be a marker for short survival in GBM (Carpenter et al., 2012). Hedgehog/GLI1 signaling promoted GBM cell migration and invasion abilities via upregulation of MMP-2 and MMP-9. However, this upregulation was shown to be PI3K/Akt pathway dependent, as shown by studies using PI3K inhibitor LY294002. Thus, there exists a crosstalk between Hedgehog signaling and PI3K/Akt pathways, and increased GBM cell migration and invasion is a concerted effect of these pathways (Chang et al., 2015).

Since Hedgehog/GLI1 signaling in GBM results in an aggressive phenotype with enhanced migration and invasive ability, the inhibition of this pathway has been considered as a therapeutic strategy for GBM treatment. Various inhibitors of the Hh pathway Glabrescione B, Erismodegib or cyclopamine derivatives show their anti-invasive effect in GBM mainly by affecting SMO or GLI levels/activity and are under preclinical or clinical trials for GBM treatment.

Like most of the cellular processes, Hedgehog signaling in GBM is also regulated by miRNAs (see Figure 9.3). Gu et al. (2014) identified Hedgehog signaling specific miRNAs by performing microarray using GBM samples divided into two groups namely GLI1 high and GLI1 low depending on GLI1 expression. Of many miRNAs, miR-144 (upregulated in GLI1 high group having positive prognosis in GBM) and miR-125b-1 (downregulated in GLI1 high group) were shown to complement and antagonize the Hedgehog signaling, respectively. However, no direct link was reported between the differentially expressed miRNAs and Hedgehog signaling components. In another study, miR-9 whose expression was increased in GBM patients and cell line (patient derived, early passage) was shown to directly target PTCH1 to increase Hedgehog signaling (Munoz et al., 2015; Ben-Hamo et al., 2016). The ectopic overexpression of miR-137, whose expression was significantly lower in GBM samples, decreased the stemness of GSCs by targeting RTVP-1 which

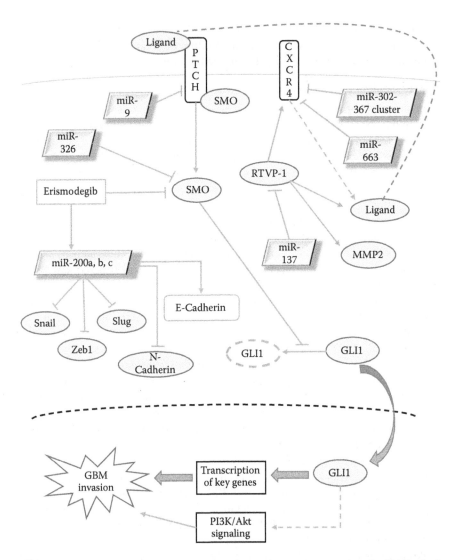

FIGURE 9.3 Interactions of Hedgehog signaling pathway components and miRNAs in the regulation of GBM invasion. Pro-invasive genes are shown in gray eclipses while anti-invasive genes are shown in rounded rectangles. The highlighted parallelograms represent miRNAs suppressing GBM invasion. Bold and dotted connections represent direct and indirect effects, respectively.

downregulates CXCR4, a positive regulator of perivascular invasion in GBM (Bier et al., 2013; Yadav et al., 2016). The levels of miR-137 in GBM patient sera were also associated with a better prognosis (Li et al., 2016). miR-302-367 cluster and miR-663 were also shown to inhibit GBM infiltration of healthy tissue via direct inhibition of CXCR4 (Fareh et al., 2012; Shi et al., 2015). miR-326, whose low levels in GBM were associated with poor prognosis, directly targeted SMO and inhibited the

SMO-dependent invasive phenotype of GBM cells (Du et al., 2015; Qiu et al., 2013; Wang et al., 2013). Overall, specific miRNAs emerge as promoters or inhibitors of the EMT and invasion phenotype in GBM by directly or indirectly affecting the wide array of factors or pathways governing this phenotype.

9.3.3 ROLE OF HYPOXIA REGULATED miRNAs IN GBM INVASION

GBM is characterized by extensive regions of hypoxia ($\sim$ 0.1-3% O_2, less than 5 to 10 mmHg O_2) that typically arises due to rapid cell proliferation, lack of blood vessels supplying oxygen to the cells deep within the hypoxic core and a high extent of necrosis that inhibits oxygen perfusion (Karsy et al., 2016). The extent of hypoxia has been correlated to enhanced motility/invasion, angiogenesis, stemness and development of resistance to chemo or radiotherapy, tumor aggressiveness and poor prognosis of GBM patients (Yang et al., 2012). Thus, it is widely agreed that hypoxia or hypoxia-regulated pathways are potential treatment targets for GBM patients.

9.3.3.1 Interplay of miRNAs and Hypoxia

Hypoxia-inducible factor (HIF) is a transcription factor that plays a central role in hypoxia signaling (Semenza, 1998). In response to hypoxia, HIF1α is stabilized, translocates to the nucleus and heterodimerizes with constitutively expressed HIF1β to transactivate various genes and miRNAs involved in adaptation to hypoxia (Kulshreshtha et al., 2007; Yang et al., 2012). The levels of several hypoxia-regulated genes such as HIF1α, GLUT1, VEGF and CAIX were found to be significantly associated with tumor grade and poor prognosis in GBM patients as compared to patients with lower grade glioma (Flynn et al., 2008). Similarly, a hypoxia signature of miRNA expression was obtained in GBM, where HIF regulated miR-210-3p (the most induced miRNA) was shown to promote GBM cell survival and impart chemoresistance in GBM. Several hypoxia-regulated miRNAs were found to be transcriptionally regulated by HIF in GBM (Agrawal et al., 2014). Interestingly, HIF was also shown to affect processing of specific miRNAs in GBM. For example, under hypoxia, HIF1α promoted the incorporation of pri-miR-215 into Drosha complex in GICs (Hu et al., 2016). miRNAs have been shown to affect HIF signaling by direct targeting of the HIF subunits (miR-182 targets HIF2α; miR-210-3p targets HIF3α, a negative regulator of HIF1α signaling), or indirectly by affecting its regulators such as VHL or PTEN (Agrawal et al., 2014; Kouri et al., 2015; Li et al., 2012; Zhang et al., 2014). Inhibition of miR-566 was shown to inhibit GBM cell invasiveness by enabling the accumulation of von Hippel-Lindau (VHL) to promote degradation of β-catenin and consequently inhibit hypoxia signaling via HIF1α (Zhang et al., 2014; Xiao et al., 2016). In contrast, miR-17-5p promoted migration and invasiveness in GBM cells by targeting PTEN, which consequently led to the upregulation of VEGF and HIF1α (Li et al., 2012). miR-9 was found to be significantly upregulated in invasive glioma cells as compared to the cells forming the bulk of the primary tumor. HIF1α that was also shown to be induced in invasive glioma cells, upregulated the expression of miR-9 to promote invasiveness (Katakowski et al., 2016). Paradoxically, miR-9 was also shown to upregulate E-Cadherin expression, which is a hallmark of non-invasive tumors. Based on these findings, it was hypothesized in the "Grow-Go-Grow" hypothesis that miR-9 expression was a function of hypoxia, and that miR-9 also inhibited invasiveness and promoted proliferation in the areas away from the primary tumor bulk where hypoxia is less prevalent. Thus, it is apparent that the role of miRNAs might be biphasic, as in this case, and a careful assessment of such a phenotypic shift (invasive to proliferative phenotypes) is essential during functional characterization of miRNAs in glioma.

9.3.4 REGULATION OF NUTRIENT UPTAKE BY miRNAs

The metabolic reprogramming in cancer cells promotes increased nutrient uptake and metabolism for increased survival, proliferation and invasion. Specific miRNAs were also shown to affect invasion by affecting the nutrient uptake. miR-326 inhibited GBM cell survival by targeting Pyruvate Kinase M2 (PKM2) to reduce metabolic activity and ATP levels in the cells, and PKM2 knockdown using siRNA was shown to inhibit GBM cell invasiveness (Kefas et al., 2010). miR-218, that was shown to inhibit GBM cell invasion (Liu et al., 2012), inhibited GBM cell growth and glucose consumption by targeting E2F2 (Zhang et al., 2015). Thus, miRNAs seem to exert their tumor suppressive effects via regulation of nutrient uptake and metabolism in GBM. However, miRNAs that regulate GBM cell metabolism could also exhibit a dual role. For instance, miR-451 is downregulated in GBM especially in the tumor core (Ansari et al., 2015). Interestingly, low miR-451 levels were associated with less tumor proliferation but enhanced tumor cell migration by bringing about less mTOR pathway activation and higher Rac1 pathway activation, respectively (Ansari et al., 2015; Zhao et al., 2017). Thus, for therapy, it is imperative to consider the dual nature of response of miRNAs with respect to the changing tumor microenvironment. Taken together, miRNAs regulating hypoxia and nutrient uptake in GBM could be potentially employed in inhibition of GBM invasiveness.

9.4   miRNA INTERVENTION—A THERAPEUTIC APPROACH IN GBM

Novel therapeutic agents need to be developed especially in the case of treating GBM owing to its diverse molecular signature comprising accumulated mutations, and deregulated pathways involving complex interplay between the regulators, effectors and miRNAs. While the current therapies may have marginally increased the GBM patient survival from less than a year to up to two years, these have been unsuccessful in defying death. The other single targeted therapies under clinical trials may have success in only a handful of patients in case of GBM, which relies on activation of multiple pathways for survival. In this regard, miRNA based therapies have been particularly appealing due to their ability to target multiple genes, and thus, reprogram biological pathways. miRNA-based functional studies and therapeutics involves inhibition of oncogenic miRNAs or overexpression of tumor suppressive miRNAs and has met with reasonable success. Following approaches are used for modulation of miRNA levels in vitro and in vivo.

9.4.1 MiRNA Inhibition Strategies

The various strategies used for miRNA inhibition involve the use of synthetic oligonucleotides [anti-miRs (modified antisense oligonucleotides complementary to mature miRNA) and miRmasks (modified oligonucleotides complementary to specific miRNA binding sites in target mRNA)], miRNA expression cassettes [miRsponge (plasmids encoding transcripts with tandem miRNA binding sites that can sequester miRNAs from endogenous targets)], or small molecular inhibitors (drugs effecting miRNA maturation and degradation) (Garzon et al., 2010). The synthetic RNA molecules are usually modified at the 2'-O residues to promote stability and increase binding affinity. The modifications include 2'O-Methylation (2'-OMe), 2'-Methoxyethyl (2'-MOE), 2'-Fluoro (2'-F) and Locked Nucleic Acid (LNA) (Stenvang et al., 2012). Each strategy has its own pros and cons. The most widely used are the antimiRs, however, miRsponges are getting attention due to their advantage of being cost-effective, stabile and long-term delivery.

9.4.2 MiRNA Overexpression Strategies

miRNA overexpression involves use of miRNA mimics (modified double-stranded RNAs that mimic endogenous miRNAs) or expression cassettes (plasmid or viral vectors overexpressing miRNA under control of a strong constitutive or inducible promoter). Although miRNA mimics have been employed in a vast majority of cases for functional analysis of miRNAs, some reports mention off-target effects of using mimics, thus, encouraging caution in their use (Jin et al., 2015).

9.4.3 MiRNA Delivery

miRNA delivery remains the major bottleneck owing to its poor in vivo stability, off-target effects, issues with organ specificity and effects on endogenous RNA machinery (Zhang et al., 2013). While use of expression cassettes can circumvent some of the concerns, viral delivery systems have issues related to safety, toxicity and immunogenicity limiting their clinical perspective (Zhang et al., 2013). Various miRNA delivery vehicles have been developed such as liposomes, polymers or their conjugates and exosome mediated (Baumann et al., 2014).

Some of the attempts for miRNA delivery in GBM have been discussed below. Anti-miR-21 oligonucleotides were encapsulated in chlorotoxin coupled lipid complexes to synthesize Stable Nucleic Acid Lipid Particles (SNALPs) to silence miR-21 in a mouse model of GBM by intravenous injections (Costa et al., 2015). Another approach of delivering miRNAs is by conjugation with nanoparticles to enhance the stability of the oligonucleotides and to ensure targeted delivery with minimal immunological response and side effects. This approach was used to silence miR-21 in a GBM mouse model using systemically injected nanoparticles with LNA-anti-miR-21 conjugated with Folate-conjugated three-way junction based RNA nanoparticles (FA-3WJ-LNA-miR-21) (Lee et al., 2017). Babae et al., (2014) also used similar strategy to systemically deliver miR-7 by its conjugation with integrin-targeted biodegradable polymer nanoparticles (Babae et al., 2014). Novel nanogels based on a polyglycerol scaffold were designed to successfully deliver active miR-34a to GBM tumors in mice (Shatsberg et al., 2016). The nanogels encapsulating gold nanoparticles conjugated with miR-218 mimics and chemotherapy drug Temozolomide were shown to exert a synergistic inhibitory effect on GBM tumors in mice by allowing a sequential release of Temozolomide followed by mir-218 mimics from the complex (Fan et al., 2015).

As opposed to systemic intravenous delivery, local delivery of drugs is performed using Convection Enhanced Delivery (CED) especially in cases where Blood-Brain Barrier (BBB) poses a hindrance in therapy. CED was used to locally deliver anti-let7a oligonucleotides as a therapeutic strategy to treat highly invasive GBM tumors in a mouse xenograft model. The study demonstrated that the method was efficient and that it maintained the integrity of anti-miR with no adverse effects on the recipient (Halle et al., 2016). To achieve a targeted delivery of oligonucleotides in GBM therapy, another approach was designed involving microvesicles. Human liver stem cell-derived microvesicles were shown to inhibit GBM cell growth in vitro by mediating transfer of tumor suppressive miRNAs (Fonsanto et al., 2012). Interestingly, mesenchymal stem cells were used to functionally deliver anti-miR-9 via exosomes to impart sensitivity to TMZ-resistant GBM cells (Munoz et al., 2013).

The previously designed approaches of miRNA delivery and the development of novel targeted delivery methods, like the ones mentioned above, open a new horizon in the field of miRNA-based therapeutics for the treatment of GBM. Such approaches need to be further refined and tested in order to achieve translational advantage.

9.5   CONCLUSIONS

The high extent of invasiveness of GBM is the reason behind the recurrence of GBM tumors and the very poor prognosis in GBM patients. Various studies focusing on GBM invasiveness have revealed the role of the EMT program and the participation of many effectors of EMT. Notably, the players that are reported to regulate such effectors, and thus EMT, could be grouped into distinct pathways like PI3K/Akt signaling, Wnt signaling and Hedgehog signaling. Inhibition of these players using novel pharmacological inhibitors has been fruitful in vitro and, in some cases, in vivo. Interestingly, miRNAs have shown potential in regulating the invasive phenotype of GBM by controlling the expression of the components of such pathways. The characterization of miRNAs into pro and anti-invasive miRNAs has unraveled their potential to be developed as a therapeutic tool. A better understanding of varying tumor microenvironment and its effect on miRNA expression could lead to the delineation of precise molecular therapy. Besides, the ability of miRNAs to target multiple pathways could be harnessed to develop a broad spectrum therapeutic approach. Combining the pharmacological pathway inhibitors with miRNA-based therapy could be employed to achieve a synergistic effect on suppression of GBM cell invasiveness. Changing intra-tumoral conditions like oxygen levels and nutrient uptake should be carefully assessed owing to their unexpected effects on pathogenesis and presumably differing response to therapy. Nevertheless, these factors could eventually aid in the development of personalized therapy. With the advent of development of novel miRNA and drug delivery strategies, the persisting problems of miRNA-based therapeutics could be overcome in order to devise an efficient treatment of GBM.

REFERENCES

Adamson, D.C., Rasheed, B.A., McLendon R.E. et al. 2010. Central nervous system. *Cancer Biomark.* **9** (1–6): 193–210.

Agrawal, R., Pandey P., Jha P. et al. 2014. Hypoxic signature of microRNAs in glioblastoma: Insights from small RNA deep sequencing. *BMC Genom.* **15**: 686.

Andreeva, A.V., Kutuzov M.A. 2010. Cadherin 13 in cancer. *Genes Chromosomes Cancer.* **49** (9): 775–90.

Ansari, K.I., Ogawa D., Rooj A.K. et al. 2015. Glucose-based regulation of miR-451/AMPK signaling depends on the OCT1 transcription factor. *Cell Rep.* **11** (6): 902–9.

Anton, K., Baehring J.M., Mayer T. 2012. Glioblastoma multiforme: Overview of current treatment and future perspectives. *Hematol Oncol Clin North Am.* **26** (4): 825–53.

Asuthkar, S., Velpula, K.K., Chetty, C. et al. 2012. Epigenetic regulation of miRNA-211 by MMP-9 governs glioma cell apoptosis, chemosensitivity and radiosensitivity. *Oncotarget.* **3** (11): 1439–54.

Babae N., Bourajjaj M., Liu Y. et al. 2014. Systemic miRNA-7 delivery inhibits tumor angiogenesis and growth in murine xenograftglioblastoma. *Oncotarget.* **5** (16): 6687–700.

Bartel D.P. 2009. MicroRNAs: Target recognition and regulatory functions. *Cell.* **136** (2): 215–33.

Baumann V., Winkler J. 2014. miRNA-based therapies: Strategies and delivery platforms for oligonucleotide and non-oligonucleotide agents. *Future Med Chem.* **6** (17): 1967–84.

Ben-Hamo R.., Zilberberg A., Cohen H. et al. 2016. hsa-miR-9 controls the mobility behavior of glioblastoma cells via regulation of MAPK14 signaling elements. *Oncotarget.* **7** (17): 23170–81.

Bhuvanalakshmi G., Arfuso F., Millward M. et al. 2015. Secreted frizzled-related protein 4 inhibits glioma stem-like cells by reversing epithelial to mesenchymal transition, inducing apoptosis and decreasing cancer stem cell properties. *PLoS One.* **10** (6): e0127517.

Bier A., Giladi N., Kronfeld N. et al. 2013. MicroRNA-137 is downregulated in glioblastoma and inhibits the stemness of glioma stem cells by targeting RTVP-1. *Oncotarget.* **4** (5): 665–76.

Brown T.J., Brennan M.C., Li M. et al. 2016. Association of the extent of resection with survival in glioblastoma: A systematic review and meta-analysis. *JAMA Oncol.* **2** (11): 1460–9.

Canel M., Serrels A., Frame M.C. et al. 2013. E-cadherin-integrin crosstalk in cancer invasion and metastasis. *J Cell Sci.* **126** (2): 393–401.

Carpenter R.L., Lo H.W. 2012. Hedgehog pathway and GLI1 isoforms in human cancer. *Discov Med.* **13** (69): 105–13.

Chang L., Zhao D., Liu H.B. et al. 2015. Activation of sonic hedgehog signaling enhances cell migration and invasion by induction of matrix metalloproteinase-2 and -9 via the phosphoinositide-3 kinase/AKT signaling pathway in glioblastoma. *Mol Med Rep.* **12** (5): 6702–10.

Chen L., Zhang K., Shi Z. et al. 2014. A lentivirus-mediated miR-23b sponge diminishes the malignant phenotype of glioma cells in vitro and in vivo. *Oncol Rep.* **31** (4): 1573–80.

Conti A., Romeo S.G., Cama A. et al. 2016. MiRNA expression profiling in human gliomas: Upregulated miR-363 increases cell survival and proliferation. *Tumour Biol.* **37** (10): 14035–48.

Costa P.M., Cardoso A.L., Custódia C. et al. 2015. MiRNA-21 silencing mediated by tumor-targeted nanoparticles combined with sunitinib: A new multimodal gene therapy approach for glioblastoma. *J Control Release.* **207**: 31–9.

Cuddapah V.A., Robel S., Watkins S. et al. 2014. A neurocentric perspective on glioma invasion. *Nat Rev Neurosci.* **15** (7): 455–65.

Delic S., Lottmann N., Stelzl A. et al. 2014. MiR-328 promotes glioma cell invasion via SFRP1-dependent Wnt-signaling activation. *Neuro Oncol.* **16** (2): 179–90.

Dong C.G., Wu W.K., Feng S.Y. et al. 2012. Co-inhibition of microRNA-10b and microRNA-21 exerts synergistic inhibition on the proliferation and invasion of human glioma cells. *Int J Oncol.* **41** (3): 1005–12.

Du R., Petritsch C., Lu K. et al. 2008. Matrix metalloproteinase-2 regulates vascular patterning and growth affecting tumor cell survival and invasion in GBM. *NeuroOncol.* **10** (3): 254–64.

Du W., Liu X., Chen L. et al. 2015. Targeting the SMO oncogene by miR-326 inhibits glioma biological behaviors and stemness. *NeuroOncol.* **17** (2): 243–53.

Duchartre Y., Kim Y.M., Kahn M. 2016. The Wnt signaling pathway in cancer. *Crit Rev Oncol Hematol.* **99**: 141–9.

Eder K., Kalman B. 2014. Molecular heterogeneity of glioblastoma and its clinical relevance. *Pathol Oncol Res.* **20** (4): 777–87.

Ehtesham M., Sarangi A., Valadez J.G. et al. 2007. Ligand-dependent activation of the hedgehog pathway in glioma progenitor cells. *Oncogene.* **26** (39): 5752–61.

Elias M.C., Tozer K.R., Silber J.R. et al. 2005. TWIST is expressed in human gliomas and promotes invasion. *Neoplasia.* **7** (9): 824–37.

Fan L., Yang Q., Tan J. et al. 2015. Dual loading miR-218 mimics and Temozolomide using AuCOOH@FA-CS drug delivery system: Promising targeted anti-tumor drug delivery system with sequential release functions. *J Exp Clin Cancer Res.* **34**: 106.

Fareh M., Turchi L., Virolle V. et al. 2012. The miR 302-367 cluster drastically affects self-renewal and infiltration properties of glioma-initiating cells through CXCR4 repression and consequent disruption of the SHH-GLI-NANOG network. *Cell Death Differ.* **19** (2): 232–44.

Feng L., Ma J., Ji H. et al. 2017. miR-330-5p suppresses glioblastoma cell proliferation and invasiveness through targeting ITGA5. *Biosci Rep.* Jun 21; **37**(3): BSR20170019.

Flynn J.R., Wang L., Gillespie D.L. et al. 2008. Hypoxia-regulated protein expression, patient characteristics, and preoperative imaging as predictors of survival in adults with glioblastomamultiforme. *Cancer.* **113** (5): 1032–42.

Fonsato V., Collino F., Herrera M.B. et al. 2012. Human liver stem cell-derived microvesicles inhibit hepatoma growth in SCID mice by delivering antitumor microRNAs. *Stem Cells.* **30** (9): 1985–98.

Fu J., Rodova M., Nanta R. et al. 2013. NPV-LDE-225 (Erismodegib) inhibits epithelial mesenchymal transition and self-renewal of glioblastoma initiating cells by regulating miR-21, miR-128, and miR-200. *NeuroOncol.* **15** (6): 691–706.

Gabriely G., Wurdinger T., Kesari S. et al. 2008. MicroRNA 21 promotes glioma invasion by targeting matrix metalloproteinase regulators. *Mol Cell Biol.* **28** (17): 5369–80.

Garzon R., Marcucci G., Croce C.M. 2010. Targeting microRNAs in cancer: Rationale, strategies and challenges. *Nat Rev Drug Discov.* **9** (10): 775–89.

Gorojankina T. 2016. Hedgehog signaling pathway: A novel model and molecular mechanisms of signal transduction. *Cell Mol Life Sci.* **73** (7): 1317–32.

Gu J.J., Zhang J.H., Chen H.J. et al. 2016. MicroRNA-130b promotes cell proliferation and invasion by inhibiting peroxisome proliferator-activated receptor-γ in human glioma cells. *Int J Mol Med.* **37** (6): 1587–93.

Gu W., Shou J., Gu S. et al. 2014. Identifying hedgehog signaling specific microRNAs in glioblastomas. *Int J Med Sci.* **11** (5): 488–93.

Guo M., Zhang X., Wang G. et al. 2015. miR-603 promotes glioma cell growth via Wnt/β-catenin pathway by inhibiting WIF1 and CTNNBIP1. *Cancer Lett.* **360** (1): 76–86.

Guo T., Feng Y., Liu Q. et al. 2014. MicroRNA-320a suppresses in GBM patients and modulates glioma cell functions by targeting IGF-1R. *Tumour Biol.* **35** (11): 11269–75.

Halle B., Marcusson E.G., Aaberg-Jessen C. et al. 2016. Convection-enhanced delivery of an anti-miR is well-tolerated, preserves anti-miR stability and causes efficient target de-repression: A proof of concept. *J Neurooncol.* **126** (1): 47–55.

Han L., Yue X., Zhou X. et al. 2012. MicroRNA-21 expression is regulated by β-catenin/ STAT3 pathway and promotes glioma cell invasion by direct targeting RECK. *CNS Neurosci Ther.* **18** (7): 573–83.

Hayes J., Peruzzi P.P., Lawler S. MicroRNAs in cancer: Biomarkers, functions and therapy. *Trends Mol Med.* **20** (8): 460–9.

Henriksen M., Johnsen K.B., Olesen P. et al. 2014. MicroRNA expression signatures and their correlation with clinicopathological features in glioblastoma multiforme. *Neuromolec Med.* **16** (3): 565–77.

Hu J., Sun T., Wang H. et al. 2016. MiR-215 Is induced post-transcriptionally via HIF–Drosha complex and mediates glioma-initiating cell adaptation to hypoxia by targeting KDM1B. *Cancer Cell.* **29** (1): 49–60.

Hulou M.M., Cho C.F., Chiocca E.A. et al. 2016. Experimental therapies: Gene therapies and oncolytic viruses. *Handb Clin Neurol.* **134**: 183–97.

Jiao Y., Li H., Liu Y. et al. 2015. Resveratrol inhibits the invasion of glioblastoma-initiating cells via down-regulation of the PI3K/Akt/NF-κB signaling pathway. *Nutrients.* **7** (6): 4383–402.

Jin H.Y., Gonzalez-Martin A., Miletic A.V. et al. 2015. Transfection of microRNA mimics should be used with caution. *Front Genet.* **6**: 340.

Kamino M., Kishida M., Kibe T. et al. 2011. Wnt-5a signaling is correlated with infiltrative activity in human glioma by inducing cellular migration and MMP-2. *Cancer Sci.* **102** (3): 540–8.

Karsy M., Guan J., Jensen R. et al. 2016. The impact of hypoxia and mesenchymal transition on glioblastoma pathogenesis and cancer stem cells regulation. *World Neurosurg.* **88**: 222–36.

Katakowski M., Charteris N., Chopp M. et al. 2016. Density-dependent regulation of glioma cell proliferation and invasion mediated by miR-9. *Cancer Microenviron.* **9** (2–3): 149–59.

Kefas B., Comeau L., Erdle N. et al. 2010. Pyruvate kinase M2 is a target of the tumor-suppressive microRNA-326 and regulates the survival of glioma cells. *NeuroOncol.* **12** (11): 1102–12.

Kouri F.M., Hurley L.A., Daniel W.L. et al. 2015. miR-182 integrates apoptosis, growth, and differentiation programs in glioblastoma. *Genes Dev.* **29** (7): 732–45.

Kulshreshtha R., Ferracin M., Wojcik S.E. et al. 2007. A microRNA signature of hypoxia. *Mol Cell Biol.* **27** (5): 1859–67.

Kuo K.L., Lieu A.S., Tsai F.J. et al. 2015. Intramedullary spinal glioblastoma metastasis from anaplastic astrocytoma of cerebellum: A case report and review of the literature. *Asian J Neurosurg.* **10** (3): 268–71.

Lakka S.S., Gondi C.S., Yanamandra N. et al. 2004. Inhibition of cathepsin B and MMP-9 gene expression in glioblastoma cell line via RNA interference reduces tumor cell invasion, tumor growth and angiogenesis. *Oncogene.* **23** (27): 4681–9.

Lakomy R., Sana J., Hankeova S. et al. 2011. MiR-195, miR-196b, miR-181c, miR-21 expression levels and O-6-methylguanine-DNA methyltransferase methylation status are associated with clinical outcome in glioblastoma patients. *Cancer Sci.* Dec; **102** (12): 2186–90.

Lamouille S., Xu J., Derynck R. 2014. Molecular mechanisms of epithelial-mesenchymal transition. *Nat Rev Mol Cell Biol.* **15** (3): 178–96.

Lan F., Yu H., Hu M. et al. 2015. miR-144-3p exerts anti-tumor effects in glioblastoma by targeting c-Met. *J Neurochem.* Oct; **135** (2): 274–86.

Lee J.K., Joo K.M., Lee J. et al. 2014. Targeting the epithelial to mesenchymal transition in glioblastoma: The emerging role of MET signaling. *Onco Targets Ther.* **7**: 1933–44.

Lee T.J., Yoo J.Y., Shu D. et al. 2017. RNA nanoparticle-based targeted therapy for glioblastoma through inhibition of oncogenic miR-21. *Mol Ther.* **pii**: S1525-0016(16) 45424-9.

Li H.Y., Li Y.M., Li Y. et al. 2016. Circulating microRNA-137 is a potential biomarker for human glioblastoma. *Eur Rev Med Pharmacol Sci.* **20** (17): 3599–604.

Li R., Gao K., Luo H. et al. 2014. Identification of intrinsic subtype-specific prognostic microRNAs in primary glioblastoma. *J Exp Clin Cancer Res.* Jan 19; **33**: 9.

Li H., Yang B.B. Stress response of glioblastoma cells mediated by miR-17-5p targeting PTEN and the passenger strand miR-17-3p targeting MDM2. *Oncotarget.* Dec; **3** (12): 1653–68.

Li Q., Cheng Q., Chen Z. et al. 2016. MicroRNA-663 inhibits the proliferation, migration and invasion of glioblastoma cells via targeting TGF-β1. *Oncol Rep.* **35** (2): 1125–34.

Li X., Wu C., Chen N. et al. 2016. PI3K/Akt/mTOR signaling pathway and targeted therapy for glioblastoma. *Oncotarget.* **7** (22): 33440–50.

Liang H., Wang R., Jin Y. et al. 2016. MiR-422a acts as a tumor suppressor in glioblastoma by targeting PIK3CA. *Am J Cancer Res.* **6** (8): 1695–707.

Liu Y., Yan W., Zhang W. et al. 2012. MiR-218 reverses high invasiveness of glioblastoma cells by targeting the oncogenic transcription factor LEF1. *Oncol Rep.* **28** (3): 1013–21.

Liu Z., Jiang Z., Huang J. et al. 2014. miR-7 inhibits glioblastoma growth by simultaneously interfering with the PI3K/ATK and Raf/MEK/ERK pathways. *Int J Oncol.* **44** (5): 1571–80.

Loftus J.C., Ross J.T., Paquette K.M. et al. 2012. miRNA expression profiling in migrating glioblastoma cells: Regulation of cell migration and invasion by miR-23b via targeting of Pyk2. *PLoS One.* 2012; **7** (6): e39818.

Louis D.N., Perry A., Reifenberger G. et al. 2016. The 2016 World Health Organization classification of tumors of the central nervous system: A summary. *Acta Neuropathol.* **131** (6): 803–20.

Lu P., Takai K., Weaver V.M. et al. 2011. Extracellular matrix degradation and remodeling in development and disease. *Cold Spring HarbPerspect Biol.* **3** (12): a005058.

Lu S., Zhu Q., Zhang Y. et al. 2015. Dual-functions of miR-373 and miR-520c by differently regulating the activities of MMP2 and MMP9. *J Cell Physiol.* **230** (8): 1862–70.

Lu Y., Xiao L., Liu Y. et al. 2015. MIR517C inhibits autophagy and the epithelial-to-mesenchymal (-like) transition phenotype in human glioblastoma through KPNA2-dependent disruption of TP53 nuclear translocation. *Autophagy.* **11** (12): 2213–32.

Ma X., Yoshimoto K., Guan Y. et al. 2012. Associations between microRNA expression and mesenchymal marker gene expression in glioblastoma. *NeuroOncol.* **14** (9): 1153–62.

Mao X.G., Hütt-Cabezas M., Orr B.A. et al. 2013. LIN28A facilitates the transformation of human neural stem cells and promotes glioblastoma tumorigenesis through a pro-invasive genetic program. *Oncotarget.* **4** (7): 1050–64.

Mariotti A., Perotti A., Sessa C. et al. 2007. N-cadherin as a therapeutic target in cancer. *Exp Opin Investig Drugs.* **16** (4): 451–65.

Moreno-Bueno G., Portillo F., Cano A. 2008. Transcriptional regulation of cell polarity in EMT and cancer. *Oncogene.* **27** (55): 6958–69.

Morris L.G., Kaufman A.M., Gong Y. et al. 2013 Recurrent somatic mutation of FAT1 in multiple human cancers leads to aberrant Wnt activation. *Nat Genet.* **45** (3): 253–61.

Mu Q., Wang L., Yu F. et al. 2015. Imp2 regulates GBM progression by activating IGF2/PI3K/Akt pathway. *Cancer Biol Ther.* **16** (4): 623–33.

Munoz J.L., Bliss S.A., Greco S.J. et al. 2013. Delivery of functional anti-mir-9 by mesenchymal stem cell-derived exosomes to glioblastoma multiforme cells conferred chemosensitivity. *Mol Ther Nucleic Acids.* **2**: e126.

Munoz J.L., Rodriguez-Cruz V., Ramkissoon S.H. et al. 2015. Temozolomide resistance in glioblastoma occurs by miRNA-9-targeted PTCH1, independent of sonic hedgehog level. *Oncotarget.* **6**(2): 1190–201.

Myung J.K., Choi S.A., Kim S.K. et al. 2014. Snail plays an oncogenic role in glioblastoma by promoting epithelial mesenchymal transition. *Int J Clin Exp Pathol.* **7** (5): 1977–87.

Nakada M., Okada Y., Yamashita J. 2003. The role of matrix metalloproteinases in glioma invasion. *Front Biosci.* **8**: e261–9.

Nan Y., Han L., Zhang A. et al. 2010. MiRNA-451 plays a role as tumor suppressor in human glioma cells. *Brain Res.* **1359**: 14–21.

Nikuseva-Martić T., Beros V., Pećina-Slaus N. et al. 2007. Genetic changes of CDH1, APC, and CTNNB1 found in human brain tumors. *Pathol Res Pract.* **203** (11): 779–87.

Patil C.G., Yi A., Elramsisy A. et al. 2012. Prognosis of patients with multifocal glioblastoma: A case-control study. *J Neurosurg.* **117** (4): 705–11.

Paw I., Carpenter R.C., Watabe K. et al. 2015. Mechanisms regulating glioma invasion. *Cancer Lett.* **362** (1): 1–7.

Qin R., Zhou J., Chen C. et al. 2014. LIN28 is involved in glioma carcinogenesis and predicts outcomes of glioblastoma multiforme patients. *PLoS One.* **9** (1): e86446.

Qiu S., Lin S., Hu D. et al. 2013. Interactions of miR-323/miR-326/miR-329 and miR-130a/miR-155/miR-210 as prognostic indicators for clinical outcome of glioblastoma patients. *J Transl Med.* **11**: 10.

Rathod S.S., Rani S.B., Khan M. et al. 2014. Tumor suppressive miRNA-34a suppresses cell proliferation and tumor growth of glioma stem cells by targeting Akt and Wnt signaling pathways. *FEBS Open Bio.* **4**: 485–95.

Roth W., Wild-Bode C., Platten M. et al. 2000. Secreted Frizzled-related proteins inhibit motility and promote growth of human malignant glioma cells. *Oncogene.* **19** (37): 4210–20.

Sánchez-Tilló E., Liu Y., de Barrios O. et al. 2012. EMT-activating transcription factors in cancer: Beyond EMT and tumor invasiveness. *Cell Mol Life Sci.* **69** (20): 3429–56.

Satelli A., Li S. 2011. Vimentin in cancer and its potential as a molecular target for cancer therapy. *Cell Mol Life Sci.* **68** (18): 3033–46.

Semenza G.L. 1995. Hypoxia-inducible factor 1: Master regulator of O2 homeostasis. *Curr Opin Genet Dev.* **8** (5): 588–94.

Shah A., Redhu R., Nadkarni T. et al. 2010. Supratentorial glioblastoma multiforme with spinal metastases. *J Craniovertebr Junction Spine.* **1** (2): 126–9.

Shatsberg Z., Zhang X., Ofek P. et al. 2016. Functionalized nanogels carrying an anticancer microRNA for glioblastoma therapy. *J Control Release.* **239**: 159–68.

Shi Z.D., Qian X.M., Zhang J.X. et al. 2014. BASI, a potent small molecular inhibitor, inhibits glioblastoma progression by targeting microRNA-mediated β-catenin signaling. *CNS Neurosci Ther.* **20** (9): 830–9.

Shi L., Wan Y., Sun G. et al. 2014. miR-125b inhibitor may enhance the invasion-prevention activity of temozolomide in glioblastoma stem cells by targeting PIAS3. *BioDrugs.* **28** (1): 41–54.

Shi Y., Chen C., Yu S.Z. et al. 2015. miR-663 suppresses oncogenic function of CXCR4 in glioblastoma. *Clin Cancer Res.* **21** (17): 4004–13.

Shi Z., Zhang J., Qian X. et al. 2013. AC1MMYR2, an inhibitor of dicer-mediated biogenesis of Oncomir miR-21, reverses epithelial-mesenchymal transition and suppresses tumor growth and progression. *Cancer Res.* **73** (17): 5519–31.

Siebzehnrubl F.A., Silver D.J., Tugertimur B. et al. 2013. The ZEB1 pathway links glioblastoma initiation, invasion and chemoresistance. *EMBO Mol Med.* **5** (8): 1196–212.

Song H., Zhang Y., Liu N. et al. 2016a. miR-92a-3p exerts various effects in glioma and glioma stem-like cells specifically targeting CDH1/β-catenin and Notch-1/Akt signaling pathways. *Int J Mol Sci.* **17** (11): 1799.

Song H., Zhang Y., Liu N. et al. 2016b. Let-7b inhibits the malignant behavior of glioma cells and glioma stem-like cells via downregulation of E2F2. *J Physiol Biochem.* **72** (4): 733–44.

Stenvang J., Petri A., Lindow M. et al. 2012. Inhibition of microRNA function by antimiR oligonucleotides. *Silence.* **3** (1): 1.

Su J., Zhang A., Shi Z. et al. 2012. MicroRNA-200a suppresses the Wnt/β-catenin signaling pathway by interacting with β-catenin. *Int J Oncol.* **40** (4): 1162–70.

Sundar S.J., Hsieh J.K., Manjila S. et al. 2014. The role of cancer stem cells in glioblastoma. *Neurosurg Focus.* **37** (6): E6.

Svoronos A.A., Engelman D.M., Slack F.J. 2016. OncomiR or tumor suppressor? The duplicity of microRNAs in cancer. *Cancer Res.* **76** (13): 3666–70.

Tai D., Wells K., Arcaroli J. et al. 2015. Targeting the WNT signaling pathway in cancer therapeutics. *Oncologist.* **20** (10): 1189–98.

Tian Y., Hao S., Ye M. et al. 2015. MicroRNAs let-7b/i suppress human glioma cell invasion and migration by targeting IKBKE directly. *Biochem Biophys Res Commun.* **458** (2): 307–12.

Valinezhad Orang A., Safaralizadeh R., Kazemzadeh-Bavili M. 2014. Mechanisms of miRNA-mediated gene regulation from common downregulation to mRNA-Specific upregulation. *Int J Genom.* **2014**: 970607.

Vasudevan S., Tong Y., Steitz JA. 2007. Switching from repression to activation: MicroRNAs can up-regulate translation. *Science.* **318** (5858): 1931–4.

Verhaak R.G., Hoadley K.A., Purdom E. et al. 2010. Integrated genomic analysis identifies clinically relevant subtypes of glioblastoma characterized by abnormalities in PDGFRA, IDH1, EGFR, and NF1. *Cancer Cell.* **17** (1): 98–110.

Wang Q., Li X., Zhu Y. et al. 2014. MicroRNA-16 suppresses epithelial-mesenchymal transition-related gene expression in human glioma. *Mol Med Rep.***10** (6): 3310–4.

Wang S., Lu S., Geng S. et al. 2013. Expression and clinical significance of microRNA-326 in human glioma miR-326 expression in glioma. *Med Oncol.* **30** (1): 373.

Wang Z., Xue Y., Wang P. et al. 2016. MiR-608 inhibits the migration and invasion of glioma stem cells by targeting macrophage migration inhibitory factor. *Oncol Rep.* **35** (5): 2733–42.

Wei F., Wang Q., Su Q. et al. 2016. miR-373 inhibits glioma Cell U251 migration and invasion by down-regulating CD44 and TGFBR2. *Cell Mol Neurobiol.* Nov; **36** (8): 1389–97.

Wu Z., Sun L., Wang H. et al. 2012. MiR-328 expression is decreased in high-grade gliomas and is associated with worse survival in primary glioblastoma. *PLoS One.* 7 (10): e47270.

Wu D.G., Wang Y.Y., Fan L.G. et al. 2011. MicroRNA-7 regulates glioblastoma cell invasion via targeting focal adhesion kinase expression. *Chin Med J (Engl).* **124** (17): 2616–21.

Xiao B., Zhou X., Ye M. et al. 2016. MicroRNA-566 modulates vascular endothelial growth factor by targeting Von Hippel-Landau in human glioblastoma in vitro and in vivo. *Mol Med Rep.* Jan; **13** (1): 379–85.

Xu J., Sun T., Hu X. 2015. microRNA-513c suppresses the proliferation of human glioblastoma cells by repressing low-density lipoprotein receptor-related protein 6. *Mol Med Rep.* **12** (3): 4403–9.

Yadav V.N., Zamler D., Baker G.J. et al. 2016. CXCR4 increases in-vivo glioma perivascular invasion, and reduces radiation induced apoptosis: A genetic knockdown study. *Oncotarget.* **7** (50): 83701–19.

Yan W., Zhang W., Sun L. et al. 2011. Identification of MMP-9 specific microRNA expression profile as potential targets of anti-invasion therapy in glioblastoma multiforme. *Brain Res.* **1411**: 108–15.

Yang L., Lin C., Wang L. et al. 2012. Hypoxia and hypoxia-inducible factors in glioblastoma-multiforme progression and therapeutic implications. *Exp Cell Res.* **318** (19): 2417–26.

Yao Y., Ma J., Xue Y. et al. 2015. MiR-449a exerts tumor-suppressive functions in human glioblastoma by targeting Myc-associated zinc-finger protein. *Mol Oncol.* **9** (3): 640–56.

Yao C., Li P., Song H. et al. 2016. CXCL12/CXCR4 axis upregulates twist to induce EMT in human glioblastoma. *Mol Neurobiol.* **53** (6): 3948–53.

Yao Y., Xue Y., Ma J. et al. 2014. MiR-330-mediated regulation of SH3GL2 expression enhances malignant behaviors of glioblastoma stem cells by activating ERK and PI3K/AKT signaling pathways. *PLoS One.* **9** (4): e95060.

Ye X., Wei W., Zhang Z. et al. 2017. Identification of microRNAs associated with glioma diagnosis and prognosis. *Oncotarget.* Apr 18; **8** (16): 26394–403.

Yi G.Z., Liu Y.W., Xiang W. et al. 2016. Akt and β-catenin contribute to TMZ resistance and EMT of MGMT negative malignant glioma cell line. *J Neurol Sci.* **367**: 101–6.

Zhang K.L., Zhou X., Han L. et al. 2014. MicroRNA-566 activates EGFR signaling and its inhibition sensitizes glioblastoma cells to nimotuzumab. *Mol Cancer.* Mar 20; **13**: 63.

Zhang Y., Han D., Wei W. et al. 2015. MiR-218 inhibited growth and metabolism of human glioblastoma cells by directly targeting E2F2. *Cell Mol Neurobiol.* **35** (8): 1165–73.

Zhang Y., Wang Z., Gemeinhart R.A. 2013. Progress in microRNA delivery. *J Control Release.* **172** (3): 962–74.

Zhang K.L., Han L., Chen L.Y. et al. 2014. Blockage of a miR-21/EGFR regulatory feedback loop augments anti-EGFR therapy in glioblastomas. *Cancer Lett.* **342** (1): 139–49.

Zhang W., Shen C., Li C. et al. 2016. miR-577 inhibits glioblastoma tumor growth via the Wnt signaling pathway. *Mol Carcinog.* **55** (5): 575–85.

Zhang W., Zhang J., Hoadley K. et al. 2012. miR-181d: A predictive glioblastoma biomarker that downregulates MGMT expression. *NeuroOncol.* **14** (6): 712–9.

Zhao D., Wang R., Fang J. et al. 2017. MiR-154 functions as a tumor suppressor in glioblastoma by targeting Wnt5a. *Mol Neurobiol.* **54** (4): 2823–30.

Zhao K., Wang L., Li T. et al. 2017. The role of miR-451 in the switching between proliferation and migration in malignant glioma cells: AMPK signaling, mTOR modulation and Rac1 activation required. *Int J Oncol.* **50** (6): 1989–99.

Index